THOMSON

COURSE TECHNOLOGY

Professional ■ Technical ■ Reference

OCT   2006

Portable Media Devices

By Dave Field

ISBN: 1-59863-169-1

Library of Congress Catalog Card Number: 2005911174

Printed in the United States of America

06 07 08 09 10 PH 10 9 8 7 6 5 4 3 2 1

Professional ■ Technical ■ Reference

Thomson Course Technology PTR, a division of Thomson Learning Inc.
25 Thomson Place
Boston, MA 02210
http://www.courseptr.com

Publisher and General Manager, Thomson Course Technology PTR:
Stacy L. Hiquet

Associate Director of Marketing:
Sarah O'Donnell

Manager of Editorial Services:
Heather Talbot

Marketing Manager:
Heather Hurley

Acquisitions Editor:
Mitzi Koontz

Marketing Coordinator:
Jordan Casey

Project Editor:
Jenny Davidson

Technical Reviewer:
Bryan Hiquet
Eric D. Grebler

PTR Editorial Services Coordinator:
Elizabeth Furbish

Interior Layout Tech:
Digital Publishing Solutions

Cover Designer:
Mike Tanamachi

Indexer:
Katherine Stimson

Proofreader:
Sara Gullion

For KiKi, Squirt, and Scooter.

Acknowledgments

I'd like to thank Amy and the girls for once again helping me find time to crank out another book.

Thanks to John, Chloe, and Sam Rutledge for assistance with iPod and Zen photos and for John's Far Cry war stories of action with the =GZR= Clan.

Thanks Mitzi, Jenny, and the unsung heroes at Course PTR for all the work that goes into making the first books in a series successful.

Thanks Laura, Renee, Stacey, and Jamie at Studio B for all your hard work this year.

Thanks Bryan, Sara, Katherine, and Elizabeth for making sure my facts (and English) came out right.

Thanks Mike for a kickin' cover!

About the Author

Dave Field is Systems Engineer for Mall of America in Bloomington, MN. In this role Dave is responsible for the design and operation of the mall's information technology infrastructure.

After hours Dave is an author, a freelance trainer, and a presenter. Dave is an expert at computing topics, networking technologies, and consumer technology. He has delivered training at Microsoft Technical Education Centers, authored books and training guides for Microsoft and Osborne/McGraw-Hill for numerous Microsoft certifications, and written consumer technology books including *How to Do Everything with Windows XP Home Networking* (Osborne) and *Fire the Phone Company* (Peachpit Press).

Table of Contents

Table of Contents

Introduction

When I was asked to write a book for the *Gadget Geek's* series, my first reaction was, who is going to be brave enough to bring one of these books up to the counter? But then I got to thinking (as geeks are wont to do)—most geeks are not overly concerned about what others think about their geekish tendencies. We dress like "fashion don'ts." We congregate at RPG (role-playing game) and *Star Trek* conventions, speaking languages such as Klingon or Elvish. We voraciously read science-fiction and fantasy books. We're so off base that we are kind of cool. We've become something of a cultural phenomenon.

In fact, our most revered literary work has recently been made into a successful movie trilogy. (I assume you've heard about *Lord of the Rings*.) We're on a pretty good roll in terms of public sentiment toward our ilk. Geek is cool! Geek is chic! Why wouldn't you want to be seen with a book like this? It's a status symbol! An icon!

As if that wasn't enough, it's also full of all the tips and tricks I could bring together in four months of late nights and weekends to help you make the most of your portable media device.

You'll get advice on which portable media devices are best for your needs. You'll learn how to find music and video entertainment online. You'll get the most current information available on digital media production and distribution methods. I'll demystify digital media codecs, music and video formats, and digital rights management technologies. I'll show you how to move music and video from one library to another, consolidating your media into a single comprehensive collection. You'll get a clear view of the pros and cons of iTunes, Windows Media Player, and other media management platforms—from the venerable Winamp to the relatively recent Yahoo! Music Engine. You'll learn to podcast, podcatch, and podlearn. You'll get a black belt in digital media conversion. You'll learn how to talk "the talk" and walk "the walk".

Bottom line: I'll make a gadget geek out of you!

As I wrote this book, the world of digital media did not sit still:

- Apple released the 5th Generation (Video) iPod and retired the iPod mini, the 4th Generation iPod, and the photo iPod.

- Creative and iriver released over a dozen new models of media player between them.

- Archos, Sony, Panasonic, Toshiba, Motorola, Nokia, Philips, Olympus, Dell, and Samsung all released new devices as well.

- Video podcasting came to iTunes.

- New versions of podcatching software were released by Ziepod, Doppler, and Juice.

I got most of these developments into the book, giving you what is the most up-to-date reference for digital media available. I also encourage you to visit the book's errata page on Course PTR's website (www.courseptr.com) for any other late breaking information I can find between publication and the next printing. You'll also find a very cool "geek speak" glossary on the website.

CHAPTER BY CHAPTER:

This book proceeds in a logical progression from introductory overviews to in-depth how-to information. Each chapter is meant to stand alone as a complete tutorial in a specific aspect of portable media production, management, or usage.

Chapter 1: Introduction to Digital Media

This chapter gives you a tour of current digital media formats. It presents the strengths and weaknesses of each format and equips you to make informed decisions when you buy or produce media. In this chapter you'll learn the difference between AVI and AAC, WAV and WMA. You'll learn about the history of digital media and digital media devices. You'll also read about some of the impending developments in digital media.

Chapter 2: Portable Media Devices

This chapter introduces various media player form factors. The introduction will include MP3/WMA players, Portable Media Centers, and devices that play portable media in addition to other primary functions (iPAQ, Palm, and other PDA devices with media player functionality). In this chapter you'll go more in depth with portable media, and will start to see the advantages and disadvantages of various formats and devices.

Chapter 3: Apple iPod Portable Media Players

This chapter will describe the available iPod Family devices. Coverage will include a description of the different devices available, including iPod, iPod shuffle, iPod nano, and the new iPod phones. I'll also introduce iTunes and show you how to perform basic media management tasks in iTunes. I'll introduce iPod accessories such as FM transmitters, voice recorders, and car docks.

Expect to know how to buy, manage, and use music and video when you've completed this chapter.

Chapter 4: A Tour of Windows XP Media Center Edition 2005

This chapter introduces you to Microsoft Windows XP Media Center Edition 2005. While in no sense a complete examination of MCE 2005, it will help you determine whether the features of MCE 2005 are required by your digital media demands. You'll learn how Media Center Edition fits into the world of digital media, and discover many exciting features of this media system. You'll learn about available accessories for Media Center Edition and will get an overview of the recording features built onto this system.

Chapter 5: Windows Media Player 10

This chapter covers the layout and operation of Media Player 10. You will learn how to operate the basic functions to record and play media. You'll use Internet radio, Internet video, and other streaming media sources. You'll learn the basics of CD ripping and burning. I'll show you how to locate and apply skins and visualizations to customize Media Player to your own preferences. I'll also show you how to sync your media to a portable media device.

Chapter 6: Buying Digital Media

This chapter helps you navigate the numerous online music stores and choose the correct media for your needs. It will also explain the potential ramifications of various format selections and the way they affect media portability and compatibility. You'll be introduced to Napster, MSN Music, iTunes, and subscription services such as f.y.e. and CinemaNow. I'll also cover digital rights management and show you how to navigate the waters between legality and common sense. You'll learn how to back up your media licenses and how to recover them in the event of a loss.

Chapter 7: Money for Nothing and the Clips Are Free

Many artists post free songs on their websites to promote their music. Other artists release their music under Creative Commons licenses for little or no cost to the consumer. This chapter shows you how to find (legal) free music. You'll meet Winamp, the world's premier free music platform. You'll get search tricks to help you find indie music and will be shown several free media directories.

Chapter 8: Ripping, Mixing, and Burning

In this chapter, you'll learn how to copy recorded music using Media Player 10, iTunes, and Winamp. You will learn about choices of recording quality and which formats are best suited to portable media players. You will also learn how to create CD-R music albums for play in media devices or standard CD players. I'll show you how to use Windows Media Player, iTunes, and Winamp to create mixes and burn them to disc. You'll learn how to use playlists to get exactly enough music to fill a disc without overflowing it.

Chapter 9: Converting Media Formats to Play on Different Systems

In this chapter we discuss the process of converting digital media from one format to another. Examples will be provided to convert Apple's iPod music format to Windows Media Audio and vice versa. This process, called "transcoding" allows you to consolidate your music into one common format that will work in all your players and devices. I'll show you the basics of media conversion. You'll get step-by-step procedures on converting iTunes to Windows Media, Windows Media to iTunes, and both to MP3. I'll also discuss options for removing digital rights management imposed by certain media players to allow you to play your music in other media players.

Chapter 10: Managing Your Music and Video Library

This chapter shows you how to manage large numbers of audio and video clips, including renaming multiple clips and organizing clips by artist/category/genre. You will also learn how to search for clips within a system. You'll get specific advice for managing a music library in iTunes, Windows Media Player, and Winamp. You'll get tips on media tags and other digital media metadata types. You'll also learn how to consolidate your music in a common format in a common location. Finally, you'll learn how to protect your media investment of time and money by making backup copies of your library and backups of your digital rights management data.

Chapter 11: Podcasts; Not Just for Ubergeeks Any More

Podcasts have become popular as a means to distribute news, music, talk radio, and other information to online audiences. This chapter shows you how to subscribe to podcasts and listen to them on iPods and other portable media devices. You'll get all the tricks of the trade to locate, subscribe, and carry these podcasts on the road. You'll find free and low-cost podcatchers, searchable directories of podcasts, and search engine tricks to locate the more elusive podcast feeds. You'll get tips on interesting new uses for podcasting and information on the use of podcasts in learning and public relations.

Chapter 12: Start Your Own Radio Station

Podcasts can be created with free or low-cost software and posted for subscribers. This chapter describes the process and walks you through the process of posting a podcast. I'll describe the required hardware, software, and hosting. I'll describe the various features of the equipment and podcasting programs, helping you decide which is best for your show and your budget. I'll walk you through the steps of creating a podcast, from script to finished product. I'll also discuss promotion and syndication. I'll introduce you to podcasting industry organizations. Finally I'll cover other digital media services such as broadcasting with streaming or on-demand media.

Chapter 13: More Than Media

Many portable media devices have the ability to store more than digital media. This chapter describes the use of media devices to transport data between computer systems, and the other uses to which your device can be put. You'll see how to configure your music device to carry data files. I'll show you how to use your device for voice recording and podcasting. You'll see accessories that can be used to expand your devices' capacity or to connect it to entertainment systems such as television or stereo. I'll show you how to use your computer or portable media device as a jukebox, and how to professionally mix music using just two iPods and a special mixer device. Finally, I'll bring it all together in one day in the life of Dave the gadget geek, to show you how one person might use digital media in a real-life scenario.

the gadget geek's guide to
Portable Media Devices

1

Introduction to Digital Media

C'mon! I thought this book was about iPods and stuff!

Relax! We'll get there soon. There are just a few things to take care of first. If you can't help yourself and want to get down and dirty with your media device, just skip tracks to Chapter 2 or 3.

<Dave pauses 5 seconds.>

Are they gone? Good. Now let's discuss a few foundational things before we move on to the device chapters. They'll be back asking about AAC files or something and then we'll laugh!

Portable media devices use digital media files to present the songs, videos, or photos you have loaded into them. It is important to develop an understanding of these media types and how they are used.

Have you ever been trying to view video on a website and wondered why you had to load QuickTime or RealPlayer before you could watch it? QuickTime and RealVideo are types of digital video. Each has its own proprietary player that you use when you view the video. Similar things happen when you view the photos on a web page. Your web browser uses a digital photo decoder to read the picture file served up by the website and to display it for you. MP3s and WMAs (we'll decode the acronyms a little later) are digital audio formats.

In this chapter I'll introduce the common digital audio, video, and photo formats and show you a few examples of each.

HISTORY OF DIGITAL MEDIA

What I'm Listening To

Title: Dragostea din tei
Artist: O-Zone
Source: iPod nano

Flash back to the earliest science-fiction movies involving computers. They made all sorts of tweets and chirps. Some spoke in strange monotone voices with interesting nasal overtones. They usually spoke proper English, enunciating syllables clearly and distinctly. They rarely played music or video, existing primarily in a utilitarian role. What the science-fiction writers did not count on, however, was the playfulness and inventiveness of legions of college students leading the charge into the digital age. These individuals, the first true geeks, were ecstatic when they were able to cause a computer's speaker to beep songs in addition to the common warning beeps.

So, say a silent thank you to those who suffered lack of sun and a diet of Cheetos to bring us the many wonderful media products and impressive media compression and presentation formats that we now have.

Digital Audio

The first "songs" played by computer systems were beeps from the system speaker. By manipulating the frequency of the system speaker, DOS programmers were able to "write" simple songs. Some of these songs (and other sounds) were designed to accompany the first computer games. While they were very likely the object of some pride (and envy), there are no documented accounts of a DOS song ever getting a geek a date.

Here is an example of "music" created using the GW-BASIC programming language's PLAY statement:

```
10 PLAY "t120 o2"
20 PLAY "eefedddgege"
30 PLAY "gfafbgfbdbbddg"
```

Embarrassed by the audiovisual superiority of the Macintosh computer, DOS/Windows programmers strove hard during the 1990s to interface sound hardware and create music production applications for PC-compatible systems (as DOS/Windows computers were called). With the release of Microsoft Windows 95, the Windows PC began to gain on the Mac, eventually offering feature parity with the Macintosh platform.

While you'll have a difficult time getting an admission from a Macintosh aficionado, the Windows systems of today can produce music of quality rivaling that of the once-dominant Mac. That said, most high-end audio production is still done on powerful Macintosh and UNIX computer systems.

Digital Video

Digital video has followed a track very similar to that of digital audio. Macintosh computers once again led the way with graphics programming and animation. By the time Windows enthusiasts were enjoying color displays, Mac users were learning to capture and manipulate video. The MacTV, released in 1993, was able to capture frames from television broadcasts and save them as PICT files (an Apple graphics format) for use in presentations.

Once again, Windows 95 proved the trigger for widespread acceptance of digital video on the PC-compatible platform. Using the recently developed AVI format, Microsoft gave users the ability to view video snippets downloaded from the Internet or computer bulletin board systems. Affordable video capture hardware became available and Windows enthusiasts began digitizing their video collections for the amusement of others.

COMMON DIGITAL MEDIA FORMATS

What I'm Listening To

Title: Runnin' Down a Dream
Artist: Tom Petty & The Heartbreakers
Source: iPod nano

Digital media uses various methods to convert the sounds of voice or music from analog format (sound waves) to bits and bytes. The method used is defined by the *codec* used to encode the media.

Codec

A portmanteau word—or blended word—for code/decode. Codecs provide the capability to encode and decode audio or video information in digital format. Codecs vary in their approach to media encoding, some providing better fidelity than others. The resultant digital media file size also varies by codec due to the differences in compression achieved by each codec.

There are several common codecs used in digital media. Violent religious debates erupt in geek circles regarding the quality and fidelity of various codecs, so think twice before you start gushing about AAC at a Windows Media conference (for instance).

Audio Formats

Digital audio is the area of digital media that is currently getting the most attention. Digital audio devices the size of a pack of playing cards can hold over 15,000 songs. The Apple iPod nano stores up to 1,000 songs and is the size of a pack of sugarless gum!

Digital audio has been encoded in many formats over the years. Some formats have stood the test of time, whereas others have been relegated to the digital scrap heap. I'll spend some time discussing the most popular formats available today.

WAV

The WAV (short for wave) format is a digital audio format created by IBM and Microsoft that has been used for years in Intel-compatible computer systems. It encodes audio information using one of a number of codecs. Most WAV codecs produce very large files and are not suitable for portable media. WAV is used extensively to produce sounds in Microsoft Windows systems.

AIFF

The Audio Interchange File Format (AIFF) is used by Apple on most Macintosh computer systems. It is similar to WAV in that it produces large files.

MP3

The first truly portable audio format, the MPEG-1 Audio Layer 3 (MP3) was designed to be a compact audio encoding format for digital video files using the MPEG-1 video encoding format. It has since gained popularity as a stand-alone audio recording format. It has been used extensively by music enthusiasts to record music from CD for sharing over the Internet. Sites such as MP3.com enable users to locate and download music encoded in this format.

AAC

Advanced Audio Coding (AAC) is the Layer 3 audio encoder used in the MPEG-4 video standard. It offers superior fidelity at lower bit rates than the MP3 format. Apple uses AAC as the primary music format in the iPod digital media players.

WMA

Windows Media Audio (WMA) is Microsoft's own digital audio format designed to compete with MP3 and AAC for digital audio. It supports several sampling rates and bit rates. This allows the producer to tailor the quality and bandwidth requirements for the specific application.

Compound Acronyms: Say a Mouthful with a Few Letters

If you think simple acronyms such as AIFF and WMA are hard to understand, try compound acronyms (acronyms containing acronyms) like MP3. Just three little letters to say "Moving Picture Experts Group Version One Audio Layer Three"!

If this doesn't bother you, try this recursive acronym: LAME Ain't an MP3 Encoder (LAME).

RealAudio

RealAudio is a low bandwidth encoding format used primarily by Internet radio stations for streaming broadcasts. It provides decent audio quality and enables multiple clients to listen to an audio stream at once.

Video Formats

Digital audio may have more experience, but digital video has been capturing a lot of attention lately. Amazing advances have been made in video compression in recent years. Movies that once required hundreds of megabytes to store can now be downloaded to your portable media center to view on the bus ride to school or work. Internet sites now routinely show video news clips and bits of speeches. Recording artists are including (sometimes hidden) videos on their CDs.

While there have been many video formats, a relative few have gained sufficient popularity to be considered mainstream.

AVI

The Audio Video Interleave (AVI) format was designed by Microsoft in 1992 for Windows computers. It remains popular due to its ability to employ a number of codecs for video encoding. Among these codecs are Indeo, some versions of RealVideo, MPEG- 4, and DivX.

MPEG/MPG/MP4

Moving Picture Experts Group (MPEG) refers to one of several formats designed by members of the MPEG working groups. MPEG-1 was the first version ratified by this organization and was the basis for much of the digital video found online today. MPEG-2 is used extensively for DVD and digital satellite broadcasting. MPEG-4 is replacing MPEG-1 and MPEG-2 in some areas and is the basis for the latest version of Apple's QuickTime video solutions.

MPEG video files may use the MPEG, MPG, and MP4 file extensions.

MOV

MOV files are produced and played by the Apple QuickTime multimedia production environment. As Apple transitions users to the MPEG-4 standard, QuickTime video files are beginning to favor the MP4 file extension over the MOV file extension.

WMV

Windows Media Video (WMV) is Microsoft's foray into moving picture respectability. Used extensively by Windows Media Player, and produced by the free Windows Movie Maker and other video production tools, WMV is making a bid to become the de facto standard for Windows video.

RealVideo

Once the leader of streaming video formats, the RealVideo format is currently being challenged in this arena by both Windows Media and Apple QuickTime. A large number of live news broadcasts and live entertainment events still use this format, now in its tenth revision.

RealVideo requires the Real Player (downloadable for free from Real.com) for playback or content.

Photo Formats

There are probably more digital graphics formats than all other digital media formats combined. Most computer users only need to know a few of these formats. Graphics designers will be more familiar with other formats used by drawing and photo-editing applications.

BMP

Short for bitmap, BMP files (see Figure 1.1) are literally a map of pixels and the colors each one represents. As such, they are very large (up to 24 bits per pixel, with millions of pixels representing some images). BMPs are used extensively as icons and images in Microsoft Windows.

Figure 1.1
This Windows BMP file uses over 7.5 megabytes to depict the image in the photo.

PICT

PICT, short for Picture, was the original graphics format for Macintosh computers. PICT was able to define both bitmapped and vector (lines, angles, and curves) graphics. Much of the superiority of the Macintosh as a graphic platform is due to its native handling of sophisticated graphics.

GIF

An acronym for Graphics Interchange Format, GIF was designed by CompuServe for the sharing and display of color graphics images (see Figure 1.2). It remains a popular format for web page graphics, second only to the JPG format.

Figure 1.2
This GIF file uses just 600 kilobytes to depict the same photo with some loss of color quality.

JPEG/JPG

The Joint Photographic Experts Group created the JPEG standard to produce compressed bitmapped graphics (see Figure 1.3). It has become the most common graphics format on the web. It produces images of lower quality than those produced by BMP and other formats, but the images are highly compressed and can be downloaded over slow Internet connections.

Figure 1.3
This JPG file uses less than 200 kilobytes to display the image with little noticeable loss of fidelity.

PNG

Portable Network Graphics (PNG) is a bitmapped graphics format designed to replace the GIF format when patent issues arose in 1995. It supports more colors and higher compression rates than the GIF format and is used in web applications where image quality is important (see Figure 1.4).

Figure 1.4
This GIF file uses nearly 3.5 megabytes to display a quality photo.

USES FOR DIGITAL MEDIA

What I'm Listening To

Title: Delirious
Artist: ZZ Top
Source: iTunes on PC

We have come far from the days of crude beeps and pixilated graphics. Digital media today is capable of rendering detailed images as large as a billboard or a movie screen. Multi-track sound systems are used to digitally produce everything from hip-hop jams to symphonic recordings.

Personal Entertainment

One of the hottest uses for digital media is in the area of personal entertainment. Your computer is a full-fledged studio, capable of producing mixes and videos with all the quality of those you see on television or hear on the radio. Your computer also operates as a digital media collection, storing recorded and downloaded songs and videos and making them available to your portable media devices for travel-size media wherever you go.

Online Streaming Entertainment

Live radio broadcasts and concert feeds can be streamed to your computer using RealVideo, QuickTime, or Windows Media. These feeds are advertised in one of many Internet media directories (examples include WindowsMedia.com and Real.com). Media companies such as NBC, CBS, Fox, and others have extensive online libraries. You can watch streaming coverage of media events as they happen, or recorded news and entertainment.

Online Recorded Entertainment

Streaming media is only the beginning. Recorded media makes up the vast majority of online media content. Media content ranging from music posted by indie bands to recorded speeches of past presidents is available from thousands of online archives. Search engines such as Yahoo! (see Figure 1.5) and Google now offer media search tools to help you find the audio or video clips you are looking for. Enjoy the bar mitzvah video of someone you don't even know, or download the latest U2 or Black-Eyed Peas song for your iPod. You are free to explore your interests, however esoteric. Stay plugged in to the beat of the world and its changing media trends.

Personal Media Libraries

It's great to find a new interesting media clip, but even better is not having to find it again. By downloading and organizing media clips, you can build your own personal media collection on your computer. Enjoy your clips again and again, build them into mixes and playlists, and take them on the road.

Tools like iTunes and Windows Media Player allow you to create digital libraries on your computer and download them to your portable media devices. These tools are absolutely free and simple to use. They are invaluable tools for those with a large media collection.

Figure 1.5

The Yahoo! video search feature.

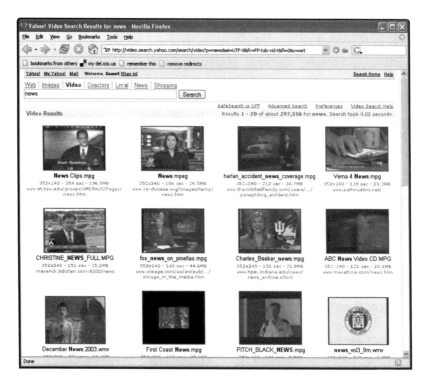

Newscasts

News, weather, and sports are another medium that has been brought online. The same media giants running the large portals have live streaming reports and recorded news available around the clock. Some, such as MSNBC (see Figure 1.6) are the online arms of larger broadcast networks, others are 100% online. Most online news organizations are digests of various news and information sources. Examples of online news digests include Slashdot.org, and the News areas of Google and Yahoo!.

News articles can be distributed directly to viewers as they browse the Internet, or they can be published for collection by smart agents on your computer.

Podcasts

Podcasts, so named because they were first designed for distribution to iPod personal digital players, are articles of news or information, even talk shows, which are "published" by their creator. They are designed to be collected by aggregator software and synchronized to your iPod.

Other personal digital players can make use of podcasts with the right software. In Chapter 11, I will show you how to subscribe to these articles using both Windows Media Player (for Windows Media devices and MP3 players) and iTunes (for iPods).

Figure 1.6
The MSNBC news portal links users to NBC's entire news organization.

Aggregator

An aggregator is a software application that can be used to subscribe to news and information updated on one or more websites. Most aggregators use a technology called syndication to retrieve updates as they are published by the site owners.

Prerecorded Media Attachments

Sound clips and snippets of video can be attached to online news articles to illustrate a point or report actual quotes from newsmakers. This multimedia presentation offers a richer experience for the reader of the article. The reporter can create animations or slideshows that inform and teach the reader about topics that are related to the story. This promotes better understanding of the news piece and helps the reader make better decisions regarding his reaction to the article. Look for these attachments as sidebars to news stories published on online news sites.

Streaming Media Broadcasts

Some news organizations stream their live broadcasts on the Internet for people to watch or listen to as events unfold. You can listen in as the Space Shuttle docks with the International Space Station, or watch as your favorite tennis player serves for the match at Wimbledon.

Communications

You can use digital media files to enhance your communications with friends, family, and the world at large. By including media clips as e-mail attachments or by placing clips in blog articles or online video sharing sites, you can offer your acquaintances the opportunity to see and hear your subject.

Blog

A weblog (or blog) is a news and information source that can be published using the facilities of an online blog site or can be constructed on your own website using blogging tools. Blogs are "read" with a web browser or a news aggregator that subscribes to the blog's syndication page, retrieving new articles as they are published.

Blogs offer personal news and information, and can be expanded to include discussions on current events and opinions.

Photo and Video Sharing

Hundreds of sites are available that offer broadband sharing facilities for a nominal fee. These sites let you upload your media content to their servers and then provide you with a URL to the content that you can attach to e-mails or blog articles. This allows you to publish broadband content without the cost of a dedicated high-speed connection.

Audio and Video E-Mail Attachments

You can also attach audio and video files directly to e-mails. While this uses more space in the recipient's mailbox, it is often the quickest and simplest way to get content to a small group of recipients.

DIGITAL MEDIA STORAGE

What I'm Listening To

Title: Since U Been Gone
Artist: Kelly Clarkson
Source: MSN Radio

Of course, there is more to digital media than its creation. There are nearly as many ways to store and transport digital media as there are types of digital media. Some storage methods lend themselves to desktop storage, whereas others are designed specifically for portability. From magnetic and optical disks spinning at thousands of revolutions per minute to electronic media components that do not move at all, digital storage devices use a number of technologies to store media.

Computer Disk Drives

Computer disk drives use magnetic disks or platters to store data in binary format. These devices range in storage capacity from the 1.44-megabyte floppy disk to the massive multi-gigabyte hard disk drives and drive systems used in desktop computers and file servers. Media stored on your home computer will most likely be stored on a hard disk drive having from several to tens of gigabytes (see Table 1.1 for an explanation of byte sizes) of available storage space. Media stored on Internet media sharing sites may be stored in large clustered storage arrays employing tens or hundreds of hard disk drives pooled together to provide terabytes (thousands of gigabytes) of storage capacity.

Table 1.1   Bytes (Mega and Otherwise)

Designation	Abbreviation	Number of Bytes
Byte	N/A	1
Kilobyte	KB	1,000
Megabyte	MB	1,000,000
Gigabyte	GB	1,000,000,000
Terabyte	TB	1,000,000,000,000

Beyond this, you don't want to know!

Compact Discs and DVDs

Compact discs and DVDs (titled digital versatile disc by some and digital video disc by others) use optical media imprinted with tiny pits that can be read by laser light to decode the disc's data. Capacities range from the 650 megabyte CD-ROM to the 17.1 gigabyte two-layer double-sided DVD. (Standard DVDs store approximately 4.7 gigabytes on a single layer media substrate.)

Compact Discs

Often referred to as compact discs (or CDs), the name Compact Disc Read Only Memory (CD-ROM) is used to denote CD media dedicated to data storage. These discs can be mass produced on stamping machines at very low cost and are excellent choices for software distribution and distribution of digital media productions. They are often used in interesting sizes for advertising purposes and mail or ship very well with minimal breakage. Being a read-only media, they are not suitable for personal data storage.

CD-R Discs

Compact Disc Recordable or CD-R and Compact Disc Re-Writable (CD-RW) are used in personal media storage because of their ability to be recorded on a home system. CD burners use lasers to heat a layer of dye to produce pits that can be read on subsequent uses. CD-RW burners have the ability to create and heal pits in the recording layer of CD-RW discs, thus writing and erasing data on these discs.

DVD and DVD-R Discs

DVD and DVD-R discs operate on the same principles as CDs, using smaller pits and tracks to compress much more data onto a single disc. Available in single-layer, double-layer, and two-sided versions of the same, these discs can store up to 17.1 gigabytes of data on a single disc.

Solid-State Media

Leaving the world of rapidly spinning media behind (I don't know about you, but I'm feeling a little dizzy), I want to show you a few things about solid-state storage technologies. Similar to the electronic memory used to store the basic start-up routines in computer systems, these devices use electronic circuits to store digital information for long periods of time without power. Recently produced flash memory devices (as they are called) can be rewritten from 10 thousand to 10 million times, depending on the circuits used on the chips. The iPod nano, using the latest generation of flash memory, stores up to 4 gigabytes of digital media within a device the size of a pack of sugarless gum, leaving room for the display, processor chip, battery, and controls (see Figure 1.7).

Compact Flash and Secure Digital Media

Compact Flash (see Figure 1.8) and Secure Digital storage cards are probably the most familiar forms of flash memory used in digital devices today. They are used extensively in digital cameras and personal media players. They range in capacity from a few megabytes to several gigabytes.

Memory Sticks

Memory Sticks, a flash memory format designed by Sony, are used with digital cameras and video recorders to store pictures and video. With similar capacities to that of compact flash cards, the choice to use one over the other is usually based on user preference rather than performance and capacity.

USB Flash Drives

With capacities rising into the several gigabyte range, USB flash drives (often called keychain drives for their sleek form factor—see Figure 1.9) are used by many to transport files and media between computer systems. Some digital media devices are adopting this form as well, the most notable device being the Apple iPod shuffle.

Figure 1.7
The iPod nano can store over 1,000 songs.

Figure 1.8
This Compact Flash (CF) card stores 4 gigabytes of data.

Figure 1.9
This USB flash drive can store 2 gigabytes of data.

DEVELOPMENTS IN DIGITAL MEDIA

What I'm Listening To

Title: How You Remind Me
Artist: Nickelback
Source: MSN Radio

As new digital media devices are created, new technological advances are incorporated into the new devices. Among these are new media formats and encoding standards, improved storage technologies, and new methods of media distribution. I will illustrate a few trends in these areas as soon as I can remember where I placed my crystal ball…

Encoding Standards

In addition to the current popular media formats, newer codecs such as Vorbis, with improved fidelity at lower bit rates, are being developed and built into the next generation of digital devices. Many iriver players now support this codec, and other manufacturers are following suit as it catches on.

Distribution Methods

In addition to the "traditional" methods of Internet distribution of digital media content, many combination phone and digital media players can download new content from the phone network, enabling users to change songs on the fly. Other trends continue to emerge as companies find additional ways to reach consumers.

New Media Devices

Apple Computers launched the iPod nano to wide acclaim in September 2005. The latest addition to the wildly popular iPod family of portable media players, the nano completes the lineup ranging from the shuffle up to the 60 gigabyte iPod, capable of storing over 15,000 songs.

Other devices are lined up on the horizon, from portable media players with expanded capacity and video capability to ultra small players likely to be lost on the way home from the store.

WHAT'S NEXT?

Okay, we've dipped our big toe into the digital murk and you're still here. Now, let's tuck our hair up under our diving cap and make a big splash! In the next chapter you'll see an array of portable media devices and be introduced to the digital media formats that play on them. You'll learn some of the history of digital media and be able to hold your own at the coffee table as you retell tales of digital empires from the misty depths of time (two or three years ago).

the gadget geek's guide to
Portable Media Devices

2

Portable Media
Devices

Well, you now know more about digital media formats than the average earbud-sporting pseudo geek. As an aspiring gadget geek, you understand the importance of not only knowing how something works, but also knowing *why* something works (the true measure of a geek).

I will introduce the array of portable media devices in this chapter. Beginning with the portable MP3 and WMA players and progressing to handheld media centers and dedicated car players, I will explore the many types of players that you can use to get the most from your media investment. If you're itching to open up your iPod and get started, by all means go ahead and meet us in Chapter 3. Those of us who are somewhat less fabulous will catch up with you shortly.

TYPES OF PORTABLE MEDIA PLAYERS

What I'm Listening To

Title: Shimmer
Artist: Fuel
Source: MSN Radio

You can find media players in many shapes and sizes today—from a tiny wristwatch to an impressive looking digital home entertainment console. Your device will express your personality and preferences regarding entertainment. Your needs (or budget) may only call for a small player capable of holding an afternoon's song list. On the other hand, you might need space for your thousands-strong music collection. I will introduce many options and let you make up your own mind about each device's suitability for your purposes.

MP3/WMA Players

The first (and still most popular) devices for digital media were the MP3 players. Rio (who unfortunately ceased operations in 2005) produced the first well-known mass market MP3 player. The Diamond Multimedia Rio PMP300 (see Figure 2.1) entered the market in the fall of 1998, to legal battles regarding music sharing.

The Rio device sparked controversy about music sharing and piracy that served only to heighten the interest in the device. Unfortunately, this legal controversy overshadowed the actual first portable MP3 device, the Eiger Labs MPMan F10/F20 (see Figure 2.2), which was released in the summer of 1998. (The point is moot, ironically, as both companies no longer manufacture media players.)

Figure 2.1
The Diamond Multimedia
Rio PMP300.

Figure 2.2
The first mass-produced
MP3 player in North
America.

Single-Format Players

Many of the older digital media players were only able to play a single digital music format. This was partially due to a limited number of available formats, MP3 being the reigning format for several years before being joined by proprietary alternatives from Microsoft and Apple and freely distributed alternatives such as the codecs of the OGG framework specification. These devices enabled users to take their digital music on the road for the first time, replacing portable CD players as the cool device to wear on one's person. Today, single-format devices are harder to come by, being replaced with devices capable of playing a number of formats. Most remaining single-format players are produced in ultra compact or even wearable forms (see Figure 2.3).

Figure 2.3
The Xonix MP3 watch plays up to eight hours of music and recharges from your computer's USB port.

Multi-Format Players

The multi-format player came later on, as engineers began adding value to the existing lines of media players. Many now play as many as six or more audio formats, from MP3 to OGG (Vorbis), WMA, ASF (a version of WMA), and even WAV. Often these devices come with increased storage and color displays for viewing photos (see Figure 2.4).

Portable Media Centers

Building on the release of Microsoft's Windows XP Media Center Edition 2005, several manufacturers have released portable media centers designed to play music and video and display photos (see Figure 2.5). These devices require Windows Media Player 10 (WMP10) or a compatible Media Center Computer to manage and synchronize content on the device. Take your home movies on the road to Grandma's house, or download several hours of movies or cartoons to occupy the kids on a cross-country train ride. Tote your entire digital photo library in the player's hard disk drive (it holds over 10,000 photos!).

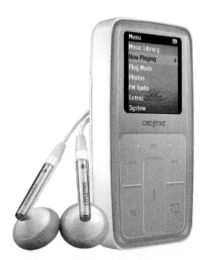

Figure 2.4
The Creative Zen MicroPhoto can play MP3, WMA, WMA-DRM, and WAV formats in addition to displaying photos.

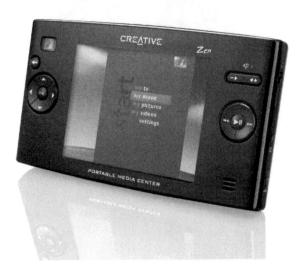

Figure 2.5
The Creative Zen Portable Media Center.

Car Audio Players

Manufacturers of high-end car audio equipment are also moving to digital audio as the preferred choice for a skip-free high quality music source. With few (or no) moving parts, MP3 and WMA players are a good choice for high quality music in the car. Some gadget geeks choose to simply use their own portable media player, connecting it into the car's audio system; others choose a dedicated player designed for the rigors of the automotive environment.

Portable Media Players with Auto Adapters

Probably the most famous example of the portable media device being connected into the car's audio environment is the iPod Beetle that was provided to those purchasing the Volkswagen Beetle in 2003. These units were engraved with the Volkswagen "Drivers Wanted" logo and came with a mix of songs and a certificate to the iTunes music store. They had their own dock designed for the Beetle that interconnected them with the car's audio system. Several manufacturers now make docking kits for automotive use with portable media devices (see Figure 2.6).

Figure 2.6
The iriver AFT 100 Car Adapter uses an FM transmitter to broadcast music to your car stereo.

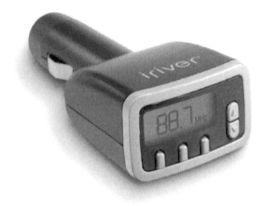

Dedicated Automotive Media Players

Digital media players designed for direct installation in cars are built to higher standards to survive the extreme temperatures and vibration of automotive life. These units resemble stereo heads complete with knobs and colorful displays. Many also play standard CDs and have AM/FM tuners. For the highest quality devices look for players from the established manufacturers in the car audio industry. Popular names such as Alpine, Kenwood, Rockford Fosgate, and others all have MP3 capabilities built into their new heads (see Figure 2.7). Most play MP3s and WMAs directly from CDs recorded on your computer.

Figure 2.7
The Pioneer DEH-P8600MP plays MP3, WMA, and WAV files in addition to CDs and AM/FM and XM Satellite radio.

Other all-digital players such as the PhatNoise PhatBox automotive media player (see Figure 2.8) provide several choices of digital media format. The PhatBox supports MP3, WMA, OGG (Vorbis), and FLAC (Free Lossless Audio Codec). It plays media from a removable hard disk drive that is loaded from a cradle attached to your computer. This unit is designed to be integrated with a high-end audio system consisting of separate amplifiers and other devices.

Figure 2.8
The PhatNoise PhatBox digital music player.

Combination Media Devices

While most portable media devices have a dedicated mission, such as playing music, some are designed to provide more than one function. Examples of this trend are personal digital assistants such as the Compaq iPAQ Pocket PC, or the Motorola ROKR iPod phone. These devices bring convergence to two major functions of electronics devices, melding form and function in new ways.

Personal Digital Assistants

PDAs (as personal digital assistants are called in geekspeak) are getting music functionality now as well. Some, like the iPAQ rx3715 (see Figure 2.9) can play music and video, take photographs, and record voice messages in addition to the common calendar/task/e-mail functions of a standard PDA.

Cell Phones

Imagine having your music fade when the phone rings. With the iPod ROKR from Motorola (see Figure 2.10), you can be jamming to your favorite tunes and the player automatically fades the music to allow you to take an incoming phone call.

Other phones provide the ability to play other formats (see Figure 2.11). Each phone supports a different combination of cellular technologies, so check with your cellular carrier to see which MP3 phones they offer.

Figure 2.9
The iPAQ rx3715.

Figure 2.10
The Motorola ROKR
phone is a fully functioning
iPod within a cell phone.

Figure 2.11
The Nokia 6630 Music
Edition phone plays music,
takes pictures and, yes,
makes phone calls.

PORTABLE MEDIA PLAYER FEATURES

What I'm Listening To

Title: Sometimes You Can't Make It On Your
Own
Artist: U2
Source: iPod nano

You've seen some of the features of portable media players as you read about the different types of players. As you can imagine, there is much more to the features and quality of a given player than what you'll see on the player's home page or the Internet. In this section I'll talk about the features you'll only see in the device's specifications listing. You'll learn how to decide which features are right for your needs and how to get right to the important facts you need to know before choosing a device.

Audio Quality

You'll see many articles and opinions about sound quality with digital media players. There are a great many variables, and almost no one takes these into account when participating in the religious discussions of whether MP3 is better than OGG or where WMA fits in. After reading this section you'll know a bit more about what goes in to good quality audio.

Sample Rate

The studio equipment that produces music CDs takes an audio snapshot of the sound being recorded a specific number of times each second. To faithfully reproduce the full range of sound that humans can hear (20Hz–20kHz), CD recording samples the audio waveforms 44,100 times each second (44.1kHz). Digital audio codecs that want to provide the best quality reproduction of CD-quality sound must approach this sample rate to avoid the perception of poor quality audio. Most common codecs can support these sample rates, some even exceed it, reaching as high as 96kHz. You'll never hear the results of audio quality that high, but your dog will!

Hertz

German physicist Heinrich Rudolf Hertz made many important contributions to the study of electromagnetism. In his honor, the term for the unit used to express frequency is named for him. One Hertz is one cycle per second and is expressed by the term 1Hz.

Bit Rate

Codecs that compress digital audio information have the ability to vary the rate of compression they use when encoding a song. By discarding sound information that is redundant and using psychoacoustics to leave out the parts of the sound that a listener is less likely to pay attention to, codecs compress audio data to lower the effective bit rate from that which would be required for CD audio. If the bit rate is lowered too much, the codec will begin to make sacrifices that are noticeable in the finished file.

Codec

The codec used to compress audio information has a lot to do with the quality of your listening experience. Each codec employs different sampling and compression techniques to encode your music. Some have more noticeable losses than others in this regard.

Lossless Codecs

Lossless codecs preserve all the music information sampled from the original source. This requires more space to store the encoded music, but results in higher quality encodings. Examples of lossless codecs include WAV, FLAC, WMA-Lossless, and Apple Lossless Audio Codec (ALAC). Each preserves all available sound information, but employs different compression techniques to store the audio data.

Lossy Codecs

You can probably already see this one coming. Lossy codecs lose some of the music information in favor of obtaining the smallest file sizes possible. Some are more successful than others at knowing which information to lose, thus debates rage regarding which loses the least amount of noticeable audio information. Current wisdom holds that the highest bit rate encodings of each popular codec retain enough information as to be unnoticeable by all but the biggest pretenders. Lossy codecs include MP3, WMA (except for WMA-Lossless), OGG (Vorbis), and AAC.

Other Factors Affecting Sound Quality

Portable music makes several compromises to justify its existence. We (correctly) assume that the ambient conditions experienced when listening to portable media will prevent us from hearing any perceptible loss of quality. We don't feel the need to dump large amounts of money on expensive equipment that will not make any noticeable improvement in the sound quality we hear.

The Player

It is interesting that while debates (and sometimes outright fights) rage regarding the quality of the codecs used, almost no one mentions the quality of the circuits that are used to implement the codec in the media player. True audiophiles (the first Gadget Geeks) know that there are large variations in quality among manufacturers of audio equipment. Folks spend entire afternoons comparing the trueness of the sound produced by one expensive component with another. Names like MacIntosh and Marantz (even Bose need not apply) are spoken of with reverence. It is little surprise then that you would be laughed out of the room if you were to enter with an iPod or a Zen music player.

Media players use circuitry designed to decode the audio information from the audio files. Different manufacturers use components designed by different circuit manufacturers for this purpose. Some reproduce the sound information more faithfully than others. The MP3 decoder in a wristwatch is likely different from one used in a home stereo system.

The Speakers

Audiophiles will spend nearly as much time selecting quality speakers as they will audio equipment. It should not surprise you that mere earbuds stuffed into your ear will offer somewhat lower quality than studio-reference headphones from the likes of Bose or Sennheiser.

Imagine my disappointment when I donned my iPod earbuds and discovered that I dance nothing like those people on the commercial!

The Room

Ambient noise will definitely affect your listening experience. It is difficult to listen to Pink Floyd when you're wondering if the infant cries are coming from your headphones or the baby in the seat across the aisle. Now, I'm not saying that you need to tell the mother to keep her kid quiet. You just need to adjust your expectations accordingly.

Storage Capacity

Storage capacity varies widely in the portable music player market. On one end, you have the 60GB music players that will hold every song you can imagine owning and a few PowerPoint presentations as well. On the other, you have the little 64MB MP3 player mounted in a pair of sunglasses. You'll load in enough songs to entertain you on a walk, but will be disappointed if you hope to load in tunes for a cross-country drive.

Hard Disk Storage

The largest storage capacities are achieved by using hard disk drives. These players might weigh slightly more, but are still surprisingly small considering how much they store. Makers like iriver, Creative, and Dell all have players in the 60+GB range with bigger units on the way. 16,000 songs might seem like a lot of music to some people, but…

Solid-State Storage

For smallest size and greatest portability, players often use solid-state (or flash memory) storage technology. These units can store as much as 8GB of music (over 2,000 songs), which is more than enough to get you across the country. If you like to keep your entire collection loaded (just in case the computer crashes), you might feel some limitations to this technology.

Format Compatibility

You have read a lot about formats and codecs, and seen the same acronyms used for both. What's up with that? Is it format or codec? Well, both. Formats are specifications that define the basic workings of a digital technology. They specify the type of encoding that will be used and provide a framework for file formatting and naming conventions. Within the format, however, is room for choice of codec. The MP3 format, for example, allows one to choose between codecs using bit rates from 16Kbps (awful) all the way up to 320Kbps (really good). Other formats specify bit rates as low as 4Kbps (WMA Voice) and as high as 1411Kbps (WAV). Your choice of a bit rate when encoding will depend on your needs for quality and storage space.

iTunes' AAC Format

Apple uses the AAC music format for selections purchased from the iTunes music service. This format can use several codec variations to produce music of various sizes and bit rates. The most common are the AAC-MPEG4 codec, which is used by Apple's iTunes and the iPod, and the AAC-LC (low complexity), which is used by some other digital audio players.

AAC is superior to MP3 in that it supports higher sampling frequencies (up to 96kHz). Sound quality as experienced using a portable media player is marginally better than that of MP3 audio using similar bit rates.

Apple AAC supports digital rights management (DRM) to protect the content of music produced by recording artists. This DRM prevents music from being copied to other devices while it is in use on a primary device. Preventing casual copying in this way improves the chances that artists will be paid for their work.

MP3 Format

The tried and true MP3 format has weathered many storms and remains a very highly used format. It has the greatest compatibility with portable players (supported by almost all devices). It can reproduce very good quality audio at higher bit rates while being able to be scaled back to store large amounts of music at lower bit rates. It is worth noting that only MP3 can be played on both the iPod and other music players. If this is an important consideration then this will be your format of choice.

WMA Format

Windows Media Audio is a format supporting dozens of codecs ranging from voice-optimized codecs at bit rates as low as 4Kbps to lossless codecs at up to 940Kbps. It is used both for portable media devices and for streaming broadcasts of Internet news and entertainment.

WMA-DRM is Microsoft's version of a digital rights management codec. It allows the user to store music on one computer and play it on any device that has been associated with that computer (but no other computers).

Other Audio Formats

Another format gaining popularity is OGG. This free format was developed to allow users to encode audio freely using "open" codecs that would never be used to charge royalties for their use. The OGG format includes more that just audio encoding. In addition to the Vorbis audio codec, the OGG specification supports the Speex speech codec, the FLAC lossless codec, the Writ text codec for titles and captions, and the Theora video codec.

Video Formats for Portable Media Players

Some portable media players are beginning to support video in the form of MotionJpeg, MPEG, and WMV (Windows Media Video). The HP iPAQ mentioned earlier can play WMA, MPEG, and MotionJpeg in addition to the audio codecs MP3 and WMA. Other devices extend the reach of Windows XP Media Center Edition 2005 as Portable Media Centers. Offering playback capabilities for MPEG and WMV, these players store as much as 60GB of digital media and photos in a player that can fit in a purse or backpack.

Additional Features

Some media players differentiate themselves by adding extra features to the mix. By offering photo displays and additional storage for computer files, these players make themselves more attractive to users who appreciate these features.

Display Technologies

With color displays as large as 3.8 inches diagonal, some players offer the ability to display movies and video in full color and sound. Feature length movies can be played on the go from movie files downloaded from services such as CinemaNow. Home movies can be loaded using WMP10 and displayed on screen or on a larger television using external connections available for this purpose.

Data Storage

Many portable media devices can make some of their storage available as computer disk storage using the computer's USB port to connect the devices as an external USB drive. This allows you to transport files between home and office, between computers in your home, or from your house to a friend's house.

Voice Recording

Some players offer the ability to record voice messages using a built-in microphone. Record memos and reminders to yourself, conduct interviews for your personal news site, or record the screams of glee that your gift of a portable music player got at your niece's birthday party.

USING YOUR PORTABLE MEDIA PLAYER

> **What I'm Listening To**
>
> Title: Love Over Gold (Live)
> Artist: Dire Straits
> Source: Zen micro

Well, I've talked the heck out of the forms and features of portable media devices. Would you like me to show you how to run one?

You'd be surprised how many folks bring home their new media player without knowing how to get music into it. They read the manual and roll their eyes when they discover that they have to go into the office and rip their entire music collection to disk before they can begin loading songs. We'll discuss ripping music in the next few chapters, and will go into great detail in

Chapter 8, but very simply, it is the process of encoding CD music in a format that your portable media player can use.

In this section I'll cover the basics of connecting and loading your media player and some tips and tricks to enjoy your music on the go.

Connecting the Portable Media Player to Your PC

Dave's Law:

To get music out, you first need to get music in.

How you get it in depends on the technology your media player uses to connect to the computer. Some connect directly using USB cables or connectors, some use wireless technologies such as Bluetooth wireless personal area networking. Others use technologies such as FireWire or flash memory cards.

USB Connections

The majority of media players can synchronize with a computer by using a USB cable or a USB connector built into the device (see Figure 2.12). These devices may include drivers that must be installed on your computer to allow it to recognize the device, or they may support a common communications interface like that used by Microsoft's PlaysForSure initiative. PlaysForSure devices can communicate with WMP10 using communications tools built into Windows Media Player.

Figure 2.12
The Creative MuVo USB 2.0 has a built-in USB connector for connection to a computer.

What Is PlaysForSure?

PlaysForSure is a Microsoft Logo program that helps consumers identify which portable media devices and online music stores are designed to work together to simplify the process of buying, loading, and playing digital media. Stores and devices using this logo (see Figure 2.13) have been certified compatible with Windows Media Player 10 and each other.

PlaysForSure music stores are certified to work with Windows Media Player 10 to allow the user to buy and download music directly into the music library stored on the computer.

PlaysForSure media players are designed to interface directly with Windows Media Player 10 to receive music selections and playlists downloaded from the computer.

PlaysForSure also applies to video players and online video stores. Stores such as CinemaNow sell feature-length movies to be downloaded to Portable Media Center devices for on-the-go viewing.

Figure 2.13
Look for the Microsoft PlaysForSure logo for assurance that devices and music stores are compatible.

Bluetooth Connections

Some media players make the connection to the computer using wireless technology such as Bluetooth (see Figure 2.14). These players can use Bluetooth to load music files as well as to function as Bluetooth headsets for cellular phones.

Figure 2.14
The Diva GEM player uses Bluetooth technology to download music from your computer wirelessly.

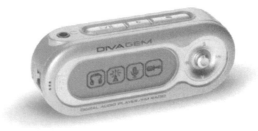

If you ask why you need to buy a portable media player that uses standard headphones just to listen to your cellular phone conversations on a headset, you're not a true Gadget Geek! Just having your conversation travel the few inches through the air from your phone to your player should be enough to justify the expense. It's like magic!

Bluetooth

Bluetooth is a specification for wireless communications among electronic items such as computers, printers, phones, personal audio devices, and digital cameras. It provides a range of approximately 30 feet and operates at speeds up to 2.1Mbps (about the speed of decent cable Internet). It is much slower than USB for equivalent applications, but offers the flexibility of not needing cables.

Other Connectivity Options

Other methods of connecting your portable media player to your computer include the use of FireWire connectivity (standard on older Apple iPods). Using flash memory media is also an option for many smaller players.

FireWire

FireWire is getting stiff competition from USB for the title of best high-speed connectivity option for computer peripherals. Less used in the Windows world, it has been the de facto standard on Apple computers for some time. The first iPods used FireWire alone to connect to computers to transfer music. Later models began using docking connectors and even (gasp) USB connectors.

No Connection

Flash memory cards are a compact means of transferring music into a portable media player. Copy your music to the memory card connected to your computer and then insert the card into your player. No muss, no fuss, no messy cables to knot and tangle!

Loading Your Music Library

Each portable media player has its own methods for loading music. Some have their own dedicated application for managing the music library on your device. Others simply let you copy files to a folder in their memory, and play it from there.

Standard Players

Standard media players may use their own application to maintain a music library for their device, or may allow you to copy music files directly to the device's memory. Applications like the Creative Media Source or iriver Music Manager manage content on older media players. These applications let you manage your music collection, build playlists, and load content into your player. The latest version of iriver's Music Manager even interfaces with WMP10 to make older players compatible with the synchronization features of WMP10.

PlaysForSure Devices

Players that work with Microsoft's PlaysForSure initiative are programmed to work with WMP10 to synchronize music with your computer. This can greatly simplify the task of loading music. Simply select the songs you want to load and add them to your Sync List. Click the Start Sync button to send music and playlists to your player.

iTunes and iPods

iTunes manages the process of loading music onto your iPod. Any song in your iTunes library is simply loaded the next time your iPod is connected. You can also define playlists that will be available to you when you use your iPod.

Enjoying Music on the Go

Most controls are pretty self evident, and are well documented on the guide included with your media player. (You DO read those don't you?) After you have loaded your player with music, there are just a few tips I can still impart.

By using preconfigured playlists and random or shuffle play features of your device, you can help your player keep you entertained with music you are in the mood for without boring you with the same sequence of songs.

Playlists

Playlists are created in your library application on your computer and loaded into your media player. Most applications allow you to include songs in one or more playlists, allowing you to keep your favorites within reach. Have one playlist for times you feel happy, and one for darker moods. Load in peppy songs for your workout or mellow ones for your afternoon under a beach umbrella.

Random or Shuffle Play Modes

Random play will help ensure you don't hear the same first five songs every time you choose a certain playlist. Often you aren't able to listen to all the songs on a list and it is nice to start in another location on the list next time you use it. Shuffling the songs lets you keep the list fresh longer.

NEXT UP, THE iPOD!

Now that you've seen a glimpse of the depth and breadth of digital media, it's time to move on to the folks who made digital media cool. Prepare yourself for a whirlwind tour of iPod features and functionality.

I'll introduce the iPod models available as this book was written (they change nearly every day), and will give you a tour of iTunes, the music store that makes your iPod work. Look for coverage of iPod accessories and references to some of the cool things you can do with iTunes and an iPod.

the **gadget geek's** guide to

Portable Media Devices

3

Apple iPod Portable Media Players

If you bought this book because you want to know more about iPods, it's payday! With just over 80% of the current portable media device market, the Apple iPod is a force to be reckoned with. Its simple navigation and clean design appeal to those who normally would be intimidated by such advanced technology.

In this chapter I will introduce the members of the iPod family, describe the features and capabilities of the iTunes music service, and give you the lowdown on iPod accessories that will make your iPod the center of the known universe for some time to come. (How's that for marketing hyperbole?)

What I'm Listening To

Title: I Zimbra
Artist: Talking Heads
Source: iPod nano

THE IPOD FAMILY OF DEVICES

The first iPod hit the streets in October of 2001. Weighing in at 6.5 ounces, it featured the smallest hard disk drive technology in a music player to date, storing 5GB of music on a 1.8-inch hard disk drive. Since then, we have seen smaller and sleeker iPods, culminating in the current iPod nano and iPod shuffle.

Note

Toshiba Corporation, maker of the first iPod hard disk drives, now manufactures hard disk drives as small as .85 inches (smaller than a quarter).

iPod

The "Big Daddy" of the lot—the iPod—is now available with 60GB of storage; enough for over 15,000 songs or 25,000 photos. Available in any color you like, as long as it's black or white, the iPod set a fashion trend with its initial stark white case and white earbud headphones (see Figure 3.1). The "silhouette people" who emblazoned busses and billboards helped bring this device to the status of a pop culture icon.

The iPod has undergone five generations of manufacturing, each slightly smaller and more refined. Current versions weigh in at fewer than 6 ounces and are slightly thinner than the first generation. Signature versions of the iPod have been released, including models featuring Madonna, Tony Hawk, Beck, No Doubt, and U2.

Figure 3.1
The iPod, now in its fifth generation (fourth generation shown), remains the flagship of the Apple fleet.

iPod shuffle

Apple's first solid-state player, the shuffle (see Figure 3.2) supports up to 1GB of flash memory (240 songs) and is controlled by a small control pad in the front of the unit. It plays songs in the order they are loaded from iTunes or in "shuffle mode" which mixes them up a bit.

Figure 3.2
The iPod shuffle may be the baby of the family, but it holds an important role as musical jewelry.

iPod nano

The iPod nano (see Figure 3.3) was introduced in September 2005 and was the second iPod (after the shuffle) to use flash memory for storage. Available in 2GB and 4GB versions, and case colors white and black, the nano impressed with its ultra small size. With 14 hours of battery life and up to 1,000 song capacity, the nano satisfied the niche for mid-size players vacated by outmoding of the mini.

Figure 3.3
The iPod nano fills the ultra-portable media player niche very nicely.

5th Generation (Video) iPod

In the midsummer of 2005, the 5th generation iPod was released. In addition to music, this unit, with the addition of video capability, plays video podcasts and television shows downloaded from iTunes. With design features very similar to the nano, this slick unit features up to 60GB of storage and can play hours of video on a single charge.

ROKR iPod Phone

Released in September 2005, the ROKR phone (see Figure 3.4) is the first phone with iPod technology built in. With 100 song capacity, it isn't the best tool for a cross-country trip, but it offers folks the chance to take a break in a hectic day of stock trading (or something) to enjoy a few tunes.

IPOD BASICS

In typical Apple fashion, the iPod was designed to be simple to operate. There are no extraneous control buttons or switches. This iconic design lends itself as much to the style of the device as it does to the function. In this section I will explore the menus and controls of the iPod. You will learn how to play music, display photos, and configure your iPod.

Figure 3.4
The ROKR phone brings
iPod music to the combo
device arena.

What I'm Listening To

Title: Prospect Hummer
Artist: Animal Collective
Source: MSN Radio

Navigating the iPod Menus

The iPod uses a hierarchical menu that branches from a simple list of options (see Figure 3.5).
Each option, in turn, offers additional choices. Choosing an option by clicking the Click Wheel
may activate that option or present you with another menu.

Figure 3.5
The iPod nano's Main
menu.

Using the Click Wheel

The iPod Click Wheel performs a number of actions in the iPod menu and for playback control. A quick glance (see Figure 3.6) shows four options arranged around a circle. Each performs the listed function when playback is underway (Play/Pause, Next Track/Fast Forward, Previous Track/Rewind). The fourth function (Menu) presents the main menu screen when pressed in playback mode, and it moves one step upward in the menu tree when it is pressed in menu mode.

If you lightly run a finger around the circle of the Click Wheel, you will notice that it performs as a quick scroll wheel in menus and as a volume control in playback mode. When a menu selection is highlighted on the screen, click the button in the center of the wheel to choose the highlighted menu selection.

Figure 3.6
The iPod Click Wheel was introduced with the iPod mini and is available on all iPods except the shuffle.

Music

The Music menu allows you to browse the iPod's music library in several ways. The playlist feature allows you to use playlists set up in iTunes or created on the go. You can also use the Music menu to browse for music by artist, genre, even composer.

Playlists

Playlists help you organize your music around your potential listening moods. You can group dance songs on your party playlist, mellow songs on your kickin' back playlist, or mix it up on your cross-country playlist.

iTunes Playlists

iTunes Playlists contain tracks that have been selected to support a certain theme or artist. They are created in iTunes and can be played in order or shuffled on your iPod. Later in the chapter, I will show you how to create playlists in iTunes.

On-the-Go Playlists

On-the-go playlists are created on the iPod itself. To add a song to the on-the-go playlist, just hold down the center click button while the song is highlighted until you see the selection flash. You will then find the song on the on-the-go playlist. The next time you connect your iPod to your computer, the new playlist will be imported into iTunes. You can then manage it as a standard playlist.

Artist, Album, Song, Genre, and Composer

Other options on the Music menu are music categories that can make it easier to locate specific songs on the iPod. They are designed to help you intuitively locate the track you are seeking.

Artist

The artist performing the track is recorded in the Artist field. This information is encoded on the CD and is detected by iTunes, or it is downloaded from the Gracenote CDDB (CD Database), an online resource for CD metadata. If more than one artist collaborated on the song you are looking for, it will be found under the primary artist's listing. When you choose an artist, that artist's albums are listed. Individual song selections can be made by choosing the album and then the song.

Album

Each CD you import into iTunes is listed in this view. Individual songs are listed under the album on which they were most recently released (usually encoded in the music file). Choosing an album displays the songs from that album that are loaded on the iPod.

Song

For a complete list of all the songs stored on your iPod, choose the Songs menu. You'd think that scrolling through 10,000 songs might be tedious… well, it is. If you rapidly circle the Click Wheel, however, it will jump through the list more quickly (advancing by page rather than song), and you will reach the bottom surprisingly fast.

Genre

As hard as they are to define sometimes, each song has a genre. Your New Age may be my Electronica, but iTunes knows for sure!

Hint: Look for Enya under New Age and Enigma under Electronica.

Composer

The original composer or songwriter is credited in this category (if known). This may be the most confusing category if you are looking for a specific song, as many of the popular songs are not written by the people you would think.

Hint: Don't look for Milli Vanilli here.

What I'm Listening To

Title: Return to Innocence
Artist: Enigma
Source: iPod nano

Photos

Beginning with the iPod photo and continuing with the iPod nano, photos can be loaded and displayed on the color screens. The tiny works of art may not be displayed in their full glory, but it is a valuable tool for backing up your photo collection in case of computer crash. You can copy your photos to any computer from your iPod to restore your collection.

Loading Photos

To load photos into your iPod, simply choose your photo folder in the iPod settings in iTunes. With your iPod connected, choose the Edit menu and select Preferences to open the iTunes Preferences window. On the iPod tab choose the Photos tab (see Figure 3.7). On this tab you can select your photo folder or specify folders to be transferred to the iPod. If you select the Include full-resolution photos checkbox, iTunes will send the full photo file to the iPod (by default, photos are scaled down for space considerations). Choose this option to have a full backup copy of your computer's photo library stored on your iPod.

Displaying Photos

Photos are displayed by navigating the Photos menu on your iPod. Each folder included within your computers photo folder will be listed as a category. Choose the folder to see a list of thumbnails of the photos within. Choosing a photo with the Click Wheel will display it full screen. Use the Click Wheel or the Next Track/Previous Track buttons to browse the photos within the folder.

The iPod photo also has a connector that can be used to display photos on television. Use this jack and the supplied adapter to view your photos on the (relatively) big screen.

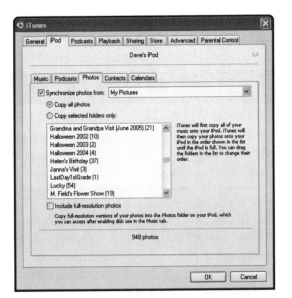

Figure 3.7
Configuring iTunes to
copy photos to an iPod
nano.

Note

The iPod nano does not include an external connector, but can still display photos and slideshows on its screen.

Slideshows

If you click the center button of the Click Wheel while a photo is displayed, you will launch a slideshow of the photos in that folder. Slideshow settings can be configured to adjust time per slide, play music during the show, change the type of transitions used between photos, even to shuffle the photos being displayed.

Extras

Included for your amusement and convenience are extra features such as a personal calendar and contact list that can be synchronized with Entourage or Outlook, a Notes tool that can be used to display text files, and games to help break up the monotony of cross-country bus rides (all while listening to music of course).

Calendar

The iPod Calendar can be used to view appointments and can be configured to beep an alarm when a meeting is approaching. Using iTunes, you can configure calendar settings on the iPod tab in the Preferences window (found on the Edit menu). Calendar applications such as Outlook can be synchronized with your iPod, and contacts can be imported from Outlook or Outlook Express.

Note

iTunes automatically synchronizes contacts and calendars from Microsoft Outlook Express and Microsoft Outlook. It does not include this ability for Entourage contacts and calendars. You can automate the synchronization of Entourage contacts and appointments by using an inexpensive third-party application called "iPod It" available from ZappTek (www.zapptek.com).

Contacts

E-mail addresses, phone numbers, and other contact information can be synchronized with the contact manager software on your computer. iTunes can automatically synchronize your contacts from Outlook and Outlook Express, or with a third-party sync manager like iPod It, you can configure synchronization with Entourage or another vCard-compatible contact manager.

Configure iTunes to Synchronize Your Outlook Contacts and Calendar

To configure iTunes to synchronize contacts and calendar items with Outlook you configure the iPod options in Preferences. Make sure your iPod is connected to your computer. Choose the Edit menu and select Preferences. On the iPod tab, choose the Contacts tab to configure contact import settings. Click the Synchronize contacts from checkbox and choose Microsoft Outlook from the drop-down menu. To synchronize calendar items, next choose the Calendars tab and check Synchronize calendars from Microsoft Outlook checkbox.

Notes

The iPod has the ability to display text notes stored on the iPod. The process for recording these notes is more complex than synching files with iTunes. You must enable the iPod to act as a removable hard disk drive and copy the notes directly to the Notes folder on the iPod (see Figure 3.8).

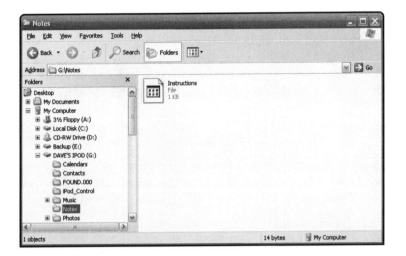

Figure 3.8
The Notes folder on an iPod connected to Windows XP as a hard disk drive.

Enable Your iPod to Act as a Hard Disk

To enable the ability to connect as a hard disk drive, you must configure the iPod in iTunes. From the Edit menu choose Preferences and select the iPod tab to display iPod options. In iPod options, click the Enable disk use checkbox (see Figure 3.9) and save the settings. The next time the iPod is connected, it will appear as a hard disk drive on your computer.

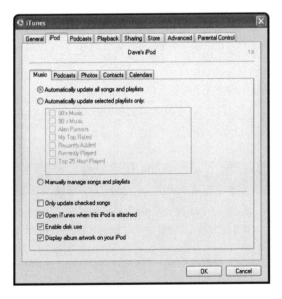

Figure 3.9
Configuring your iPod to act as a hard disk drive.

Games

The latest versions of the iPod ship with games designed to help kill the time spent by millions of iPod users who would otherwise have to stare at other riders on the train. Now, don't go thinking that the iPod is going to replace your GameBoy Advance any time soon. With titles like Parachute, Solitaire, and Brick, there is a ways to go before the iPod is raved about in gamers' magazines.

Settings

The Configuration menus on your iPod allow you to customize basic operational settings on the device. The behavior of menus, displays, and features such as shuffle and repeat are all accessed on the Settings menu. They are all pretty self-evident, but I will describe the most important ones here.

Main Menu

You can configure additional menus to appear directly on the Main menu on your iPod. Some menus normally nested under the default Main menu options can be placed directly on the Main menu by changing their status to On. If you find yourself using a particular menu frequently, you can move it up to the Main menu to save yourself time and clicks.

Equalizer

The iPod includes a built-in equalizer to adjust music according to preset EQ settings. Choose from Rock, Dance, Hip Hop, and others. EQ is found under the Settings menu.

Backlight

The backlight on your iPod can be configured to any interval from Off to Always On with stops at 2, 5, 10, 15, and 20 seconds.

Language

If you prefer a language other than English for your iPod menus, you can select from over a dozen options including Finnish (Suomi), Dutch (Nederlands), and others.

Meltdown

A word to the wise: Don't change the interface language unless you think you can find the word "language" in the language you have chosen. Luckily (if you speak English), the Reset All Settings option is always in English.

THE ITUNES MUSIC SERVICE

What I'm Listening To

Title: Immigrant Song
Artist: Led Zeppelin
Source: iTunes on PC

Without iTunes your iPod is mostly harmless. It makes an elegant paperweight and looks cool in all circumstances. No one will ever know you aren't listening to music!

iTunes loads the software necessary for your computer to communicate with your iPod. There are other third-party applications that can communicate with the iPod, but only iTunes is supported by Apple. If you choose another program such as Ephpod, Senuti, or Xpod, you are taking the risk that you'll have to fix any problems that arise yourself.

iTunes, the Media Application

iTunes is part desktop music library and part Internet music store. I will explore each portion of iTunes in turn, beginning with the desktop music application.

Ripping CDs

If you are like most new digital music enthusiasts, you'll have several dozen CDs that you'd like to get inside your iPod. The process of recording your CD music in digital format is called "ripping." We will cover this in much more detail in Chapter 8, but you'll need a basic knowledge of ripping to get started with your iPod.

Basic Ripping

When you insert a CD while iTunes is running, you will be presented with a view of the tracks on the CD. By default, all tracks will be selected and you can simply click the Import CD button located near the top right-hand corner of the window (see Figure 3.10). iTunes will record the tracks and add them to your music library. They will also be added to any smart playlists whose criteria they meet.

Figure 3.10

Recording (ripping) a CD in iTunes.

What I'm Listening To

Song: Have a Cigar
Artist: Pink Floyd
Source: iTunes on PC

Joining Tracks

You happen to love Pink Floyd and hate the way the tracks are ripped apart when they are copied to digital format. Is there some way to stitch them back together?

Better than that, you can rip them that way! Select the tracks you'd like to join—in the case of *Wish You Were Here*, all of them—and select Join CD Tracks from the Advanced menu (see Figure 3.11). Now, when you rip the disk you will get one continuous track.

Organizing Your Library

iTunes pretty much puts all your songs in one big list. You can sort them by Artist, Genre, Album, or other available criteria, but you still have a big list of songs. To make this task easier, use the Browser. This feature is activated in the Edit menu by choosing Show Browser.

The Browser (see Figure 3.12) presents a hierarchical menu for each of the Genre, Artist, and Album categories. Beginning with Genre, when you make a selection, it will narrow the selections in the remaining two menus to only those Artists or Albums that meet the criteria you have selected.

Figure 3.11
You can join tracks to pre-serve the continuity of the album during ripping.

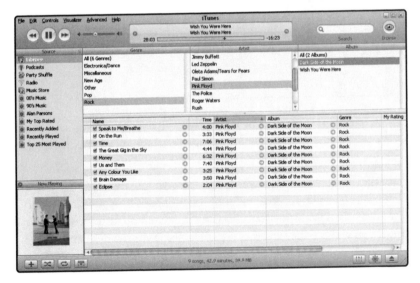

Figure 3.12
The Browser makes it eas-ier to locate tracks.

Building Playlists

Playlists do not add any organization of their own to your library, but they do allow you to play specific tracks in a specific order. You can include tracks from any album and play them in any sequence, or even shuffle them to keep the playlist fresh. When you update your iPod, the playlists you create will be sent to the iPod so you can enjoy your playlist on-the-go as well.

Standard Playlists

Standard playlists are manually configured and manually maintained. Create a standard playlist by clicking File and choosing the New Playlist menu option. Name the playlist and press Enter to save the name. To add a song to a standard playlist, simply drag it from the iTunes Library and drop it onto the playlist.

Smart Playlists

Smart playlists are designed to automatically include songs meeting certain criteria, such as Artist or Genre. They group music as it is added to iTunes and automatically update your iPod when it is connected to your computer.

Creating a Smart Playlist

In iTunes, use the New Smart Playlist option on the File menu to launch the Smart Playlist design window.

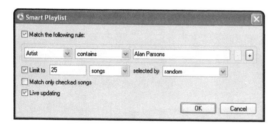

Choose query options to define the selection criteria for your playlist. Choose from Artist, Composer, Genre, or any of over a dozen other criteria. Click the + sign to the right of the criteria to add another criteria to the playlist.

It is possible to define a smart playlist that includes every song from an artist except the one or two you wish they never released. You can limit the number of songs chosen for your playlist and choose to have the playlist update itself with new selections as they are added to iTunes (live updating). When you click OK you will see the new playlist in iTunes. Type a new name for the playlist if the name iTunes chose is not to your liking.

The new playlist will appear on your iPod after the next update from iTunes.

Interfacing with Your iPod

When you have all your playlists ready—if not before—you can connect your iPod. iTunes will automatically update your iPod with any tracks you have recently added and will load in any new playlists.

iTunes, the Storefront

In addition to managing your music library, iTunes interfaces to one of the largest online music sites, namely iTunes. The iTunes site allows you to buy as many musical selections as you want for 99¢ each. The iTunes store is launched from a link on the Apple.com website, or by choosing the Music Store option in the iTunes application.

Browsing Music

iTunes is basically just a music shopping site. You can browse the site's selections to see if you can find anything interesting or use the Search tool to locate specific tracks or albums. A quick search for "droplet"—an indie rock band from Minneapolis—returns one of their albums as well as other entries probably containing the word "droplet" in a song title or album description (see Figure 3.13).

Figure 3.13
Searching for a specific artist in iTunes.

Buying Music

If you choose to buy a track (or an album), just click on the Buy Song or the Buy Album button. iTunes will ask for your account information. If you do not have an iTunes account, you can set one up in just a few minutes. When you make your selection, a confirmation window appears to verify that you intended to buy the music (see Figure 3.14). When you click Buy, the music will be charged to your credit card and the download will begin.

Figure 3.14
Verifying your iTunes purchase.

IPOD ACCESSORIES

What I'm Listening To

Title: Soft Focus
Artist: droplet
Source: iTunes on PC

Last but not least, iPod accessories are a booming market for designers of consumer electronics equipment. Makers like Bose, Belkin, and others manufacture devices such as voice recorders, cameras, external speakers, and even FM transmitters to let you make short-range broadcasts to a car or home stereo.

Voice Recorders

Plug in one of the available iPod voice recorders and your iPod becomes a personal digital recorder for recording voice memos, lectures, and interviews. Models like the Griffin iTalk (see Figure 3.15) are designed for the full-size iPod and iPod mini. The iPod nano and iPod shuffle do not support voice recording.

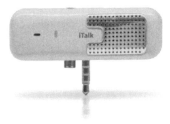

Figure 3.15
Record conversations
with a voice recorder for
the iPod.

External Speakers

Several manufacturers have designed speaker docks for the iPod. Probably most popular among these is the Bose SoundDock (see Figure 3.16). This unit actually controls the iPod with a mini remote control that comes with the unit.

Figure 3.16
The Bose SoundDock can
enable your iPod to fill the
room with noise.

Car Docks

To take your tunes on the road, you can use a device such as the Griffin RoadTrip iPod dock (see Figure 3.17). This device plugs into your car's 12V accessory plug and transmits the iPod's sound to your car stereo. It can also be used at home to transmit songs from your PC to any FM stereo.

FM Transmitters

FM transmitters like this Belkin TuneCast II Mobile FM Transmitter (see Figure 3.18) allow you to carry around your own mini radio station. Able to transmit on any open FM-band frequency between 88.1 and 107.9 MHz, it can be picked up by radios up to 30 feet away. Put your frequency on a bumper sticker and see how many friends you can pick up on the road!

Figure 3.17
Take your tunes in the car
with the Griffin RoadTrip.

Figure 3.18
Be a mini radio station with
an FM transmitter.

Protect Your iPod

A wide variety of cases and wallets are available for iPods of all sizes. From the leather iPod wallet available from Belkin (see Figure 3.19) to the iSkin mini silicone protector for the iPod mini, there are many choices for carrying and protecting your iPod.

My own personal favorite? The Louis Vuitton iPod case (see Figure 3.20). At only $215 it allows you to be seen with your iPod at charity functions and polo matches.

WHAT'S NEXT?

Well, we've toured the features and accessories available for portable media players and iPods. It's time to get an idea how these devices work with computers. We'll begin in the next chapter with Windows Media Center Edition 2005, Microsoft's operating system for media users. We'll

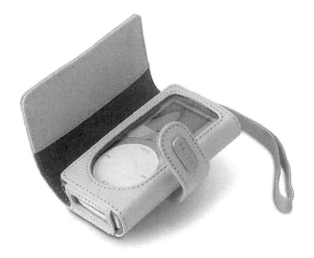

Figure 3.19
Belkin makes this fashion-
able iPod wallet.

Figure 3.20
The Louis Vuitton iPod
case is your ticket to high-
society digital music.

get in deep with Windows Media Player 10 in the next chapter, and learn how to manage and purchase media in the following chapters. You'll learn how to convert media formats, deal with digital rights management—music industry's copy prevention technology. You'll learn how to create and manage a media library to make the most of your investment in media and media devices.

the gadget geek's guide to

Portable Media Devices

4

A Tour of Windows XP
Media Center Edition
2005

In this chapter, I will briefly introduce Windows XP Media Center Edition 2005. I will tour the major functions of this operating system, and show you how it can be used to work with your digital media.

WHAT IS WINDOWS XP MEDIA CENTER EDITION?

> **What I'm Listening To**
>
> Title: Tonight, Tonight
> Artist: The Smashing Pumpkins
> Source: Windows Media Player 10

Windows XP Media Center Edition (also known as MCE) is really just Microsoft Windows XP with several extension technologies designed to enable users to manage media content more easily. This variant of the Windows XP operating system is designed for home media center operation and to be used in apartments and dorm rooms as a general purpose computing and entertainment system. It is installed as part of a special version of Windows XP and includes several menu options that are designed to be operated using a wireless remote control. The menus offer access to pictures, video, live television programming, even Internet and FM radio.

Users of MCE have access to the latest Microsoft digital media technology. Windows Media Player 10 can be found here, as can Movie Maker 2.1 and new CD/DVD creation utilities. With the ability to view, record, and pause standard and high-definition television, this system helps you make the most of your viewing time. You can use the built-in menus to browse and select upcoming programs to record, even setting recording plans to capture all scheduled upcoming episodes of your favorite shows. With MCE you get everything but the muzzle to keep your buddies from spilling the score of the game you just recorded at home.

Features of Windows XP MCE 2005

Time for a quick tour of MCE! I'll spend a few moments exploring the menus and highlighting the major features before we move on to system requirements and accessories later in the chapter. In coming chapters I will elaborate on certain technologies such as Windows Media Player 10 that are available in both MCE 2005 and separately in Windows XP.

Standard Windows Features

MCE is truly Windows XP under the covers. When installed as a desktop media center, you can close the Media Center component and use it as a normal computer. You can run Microsoft Office, computer games, surf the web—everything you'd use a regular Windows PC for. Version-wise, the flavor of Windows XP included in MCE is set someplace between Windows XP Home and Windows XP Professional. It is updated to include sexier visual elements such as the Energy Blue desktop color theme (see Figure 4.1) and blends nicely with the new versions of Media Player and Movie Maker.

Figure 4.1
Windows XP simmers just below the surface of Media Center Edition.

Media Center Features

MCE includes several features designed to enable you to enjoy digital media. Tools such as Windows Audio Converter help you convert digital audio between formats. Movie Maker 2.1 lets you create your own sophisticated home video productions. Fun features such as Windows Dancer and Party Mode (see Figure 4.2) entertain you and your guests.

Windows Media Player 10 offers excellent resources for managing digital audio and video. You can browse and buy digital music from over a half-dozen online music stores, listen to and download podcasts, and listen to live Internet radio.

Several manufacturers of television tuners offer MCE-compatible tuners for both standard and HD television. With the use of two tuners, you can even watch one program while recording another.

Extender devices allow you to access MCE features from televisions in other parts of your home. Imagine if you will, a central entertainment hub receiving, storing, and broadcasting digital media throughout your home.

Figure 4.2
Windows Party Mode lets you operate your computer as a jukebox.

Uses for Windows XP MCE 2005

Sure, Dave. Sounds great. But, what makes it worth the extra money? Can't I do all this on a regular Windows XP system?

Short answer? Yes, you can. Windows XP is after all the underlying technology that makes all this happen. The trick is making it all work together in an easy to use format so that the user can access all the sophisticated features of the system. MCE brings it all together in the menu system, allowing those with somewhat less geekiness than ourselves to be able to enjoy the power of this medium. Plus, with the wireless Media Center remote control (see Figure 4.3), you can get Cheetos dust on the couch instead of in your keyboard.

Home Media Computer

The most obvious use for Windows XP MCE 2005 would be for a computer that is used both for normal tasks and media-related tasks. Having—as it does—a full version of Windows XP under the covers, it can be used for common tasks such as paying bills, playing games, taking online courses; the list goes on.

Just click the green Media Center icon (or press the green button on your Media Center remote control) to transform your system into full media mode (see Figure 4.4). Watch television, rip CDs, download podcasts; you name it. In Media Center mode, the Windows Media wireless remote control allows you to navigate the MCE functions efficiently from across the room.

Figure 4.3
The Media Center Remote Control lets you operate MCE from your chair.

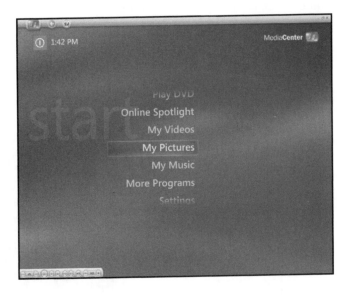

Figure 4.4
Windows Media Center Edition in Media Center mode.

Entertainment Center

Some MCE owners decide to dedicate their MCE computer to home entertainment center use. Many manufacturers design sleek MCE systems to look right at home in your entertainment center (see Figure 4.5). These systems interface with your other stereo equipment, organizing your media center into a cohesive entertainment system. With HDTV tuner cards, your MCE computer can receive and record HD broadcasts and relay them to your television. Watch the big game in HD, see who is calling when the phone rings (through available integrated

Caller-ID on-screen display), pause the video mid-play to answer the phone, rub the score in a bit, and then resume play without missing a beat.

Figure 4.5
The Origen AE X15e Home Theater PC case is designed to blend with home audio equipment.

Media Production System

As a basis for a media production system, Windows XP MCE offers beginners a great set of tools and technologies to help them become familiar with digital media production and management. With tools like Movie Maker (see Figure 4.6), and Audio Converter, tasks such as audio conversion and movie making become easier. Experience with these tools eases the transition to more powerful solutions when the need arises.

Figure 4.6
Add a variety of video effects to your home movies with Movie Maker.

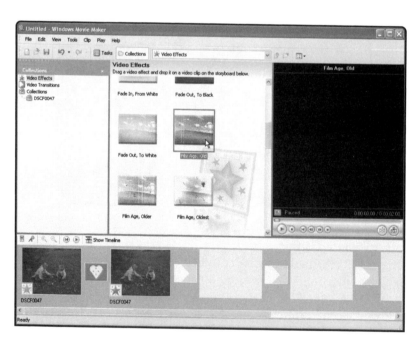

Note

If Audio Converter is not installed on your Windows XP MCE 2005 system, you can add it using the Add or Remove Programs icon in the Control Panel. Look in Add/Remove Windows Components and add Windows Digital Media Enhancements. You will need the MCE 2005 product CD to accomplish this.

GETTING THE MOST FROM WINDOWS XP MCE 2005

What I'm Listening To

Title: NPR: Story of the Day (podcast)
Artist: National Public Radio
Source: iPod nano

While this book is not intended to serve as a source of comprehensive information about Windows XP MCE 2005, there are a few tips I would like to share to help you select the right media center PC for your needs and to help get the most from your new system.

Note

Windows XP MCE 2005 is usually only available on dedicated Media Center PCs or as an upgrade for selected MCE 2004 PCs. That said, Microsoft's licensing rules allow the operating system to be sold with any compatible hardware. A quick browse to sites like pcalchemy.com will let you shop for TV tuner cards and home theater PC (HTPC) components. You will be offered the opportunity to purchase Windows XP MCE 2005 when you make any of these qualifying purchases.

Windows XP MCE 2005 Hardware

Microsoft has published minimum requirements for computers that will be used to run Windows XP MCE 2005. As you shop for systems, make sure they meet or (hopefully) exceed these minimum requirements.

Processor

In many instances, a media center PC will use the system's processor for operations requiring encoding or decoding of digital media. Digital media codecs require intense mathematical calculations in order to compress or decompress audio or video data. The minimum recommended processor for MCE 2005 is a Pentium 4 (or equivalent) processor running at 2.4GHz, with the new dual core processors from AMD and Intel being greatly preferred due to their ability to perform calculations on two different execution threads at once. In fact, a dual core processor—such as the dual core Intel Xeon—that uses hyperthreading technology can support the equivalent of four threads of execution at once!

Intel has also recently released the Viiv processor, a dual core chip based on the low-power Pentium M notebook computer processor, which is optimized for digital media processing. These processors are ideal for a system of this type.

Memory

Memory is like real estate. Build it and they will come! While 1 or 2GB of system memory may seem like a lot right now, it will seem scant enough when you are rendering a movie or converting a large audio file. As system requirements increase, memory remains the least expensive way to add performance to your system. A system with 2–4GB of RAM will serve you well into the future.

Storage

Digital media files, especially "works in progress," take up a lot of disk space. It is possible to have several versions of a movie in storage at once, along with clips and segments that have been incorporated into the production. Add digital audio tracks for the soundtrack and you can have some fairly hefty storage requirements. As a result, Media Center systems with over 500GB of disk space are fairly common.

Video

Windows MCE requires a fast video card (128 bits if possible) with at least 128MB of video RAM for optimal performance. Many video cards support onboard processing of digital video data to reduce the load on the system processor. Many features of the latest media files and many television tuner cards require DirectX 9 support as well. Check the specifications of your video card to ensure compatibility.

Hardware

If you hope to get the most from your MCE system, you'll want one or two television tuners to enable television viewing and recording capabilities, video capture equipment to support conversion of video tape libraries to digital format, and a wireless remote control or keyboard/mouse combo to allow use from a couch or chair.

Many MCE systems provide these features and more, some even including additional digital media production applications to sweeten the deal. Expect to see some systems include third-party video production software, additional storage equipment, even special keyboards and remote controls.

Extending Windows XP MCE 2005

There are several extending technologies available for Windows XP MCE to add value to an installation. Some, like Media Center Extenders, allow media to be shared with users in other rooms of the home. Peripherals such as video capture devices and elaborate speaker systems can add to your enjoyment of the system as well.

Media Center Extenders

Windows Media Center Extenders (see Figure 4.7) use your home's wired or wireless network to pass information between the media computer and a remote television. Extenders may even be an Xbox that has been extended to support this function.

Figure 4.7
The Linksys Media Center Extender uses wireless networking to expand the MCE system to other rooms in the home.

Note

The Xbox 360 includes Windows Media Center Extender functionality by default. For more information on this and other features of this gaming system, see the Course PTR title, *The Gadget Geeks Guide to Your Xbox 360.*

Media Center Peripherals

Peripherals and hardware accessories help differentiate systems. Accessories such as HDTV adapters (see Figure 4.8) enable the system to receive and process higher quality digital media. One or more of these adapters can be used to record or play video from digital broadcast stations.

Figure 4.8
The ATi HDTV Wonder digital television adapter enables a MCE system to receive up to two channels of digital HDTV broadcast simultaneously.

Portable Media Centers

A brainstorm of Microsoft's, the portable media center (see Figure 4.9) uses Windows Mobile technology to enable the user to listen to music or watch video on the go. Synchronized with the Windows Media Center console or Media Player 10, it can carry hours of video and days of music on the road.

Figure 4.9
The Creative Zen portable media center.

Using Windows XP MCE 2005

What follows is a quick overview of the operation of Windows XP MCE 2005. I'll cover how to play and record television programs, create a video DVD, and use the MSN Remote Record add-in to schedule a recording from school or the office.

Setting Up the Program Guide

Before you can use the television listing feature of the Program Guide, you must configure it to know where you are. This is accomplished in the TV Settings menu. Use the included Windows Media Center remote control to perform the following steps:

1. On the My TV menu, select Settings and then TV to open the TV Settings menu.

2. Select Set Up TV Signal to set the signal source you will be using (see Figure 4.10).

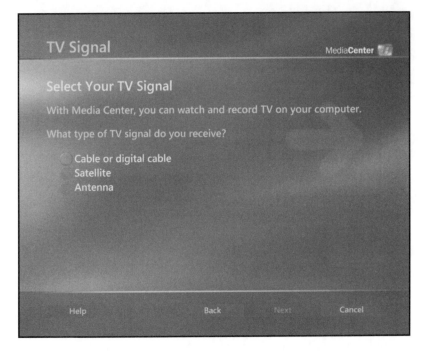

Figure 4.10
You can select from cable, satellite, and broadcast providers for the Program Guide.

3. Back on the TV Settings menu, select Guide to set Program Guide location and download listings (see Figure 4.11).

4. Once listings are downloaded, you can start viewing and recording live television.

Figure 4.11
You can use the Settings menu to configure many MCE settings, including the Program Guide.

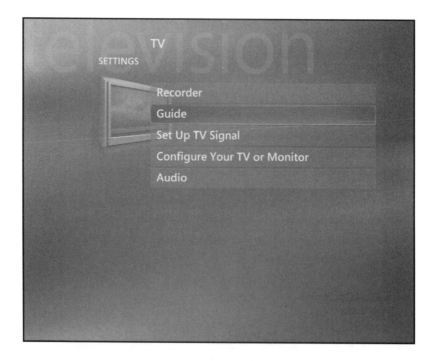

> **Note**
>
> Many TV cards require the addition of a DVD decoder to show television signals. If your Media Center computer does not include one you can download an inexpensive one from the Internet. Decoders from CyberDVD, Nero (Sonic), and nVidia do a great job.

Scheduling TV Recordings

Recording a currently playing show is pretty simple. Just press the Record button. If you want to schedule a recording, you only need to find it in the Program Guide. Once you locate your television program, highlight it and press Record. If the program you are recording is a series, press the Record button a second time to schedule the entire series for recording. Be the first on your block to have the entire season of your favorite program on DVD (coming up next)!

Copying Recorded TV to DVD

Okay! You've got your favorite show recorded in MCE 2005. You'd like to have it on DVD to take up to the cabin or on the road. You can use the built-in DVD burning capabilities of MCE 2005 to copy the program to disc.

1. Choose the My TV menu and select Recorded TV to find your recorded programs.

2. Highlight the show you want to copy and press the More Info button on your MCE remote control.

3. Choose the Create CD/DVD option to begin the process. Select Video DVD when prompted.

4. Enter a name for your recording using the remote control keypad. Use the on-screen prompts to enter the title text.

5. Press OK and then Create DVD to initiate the copy.

When the copy is done, the disc will eject. Use the disc in any DVD player.

Recording While You're Away

Microsoft had anticipated the time when you'd slap your forehead and exclaim, "Dang! I forgot to schedule the recording for "Days of Our Lives" this morning!" To help you avoid missing the show, they've created an extension called MSN Remote Recording. This tool loads on your MCE 2005 system and allows you to log in remotely to schedule a recording from school or the office.

Note

The MSN Remote Recording service is free, but requires you to use a (Free) MSN Passport account for access and also requires your MCE 2005 system to be connected with an always-on Internet connection such as cable or DSL Internet.

1. On the MSN TV listings site (http://tv.msn.com/tv/guide/), choose the show you want to record (see Figure 4.12).

2. On the Program Details page, select Record This Show (see Figure 4.13) to schedule the recording. MCE 2005 will be contacted by the application to schedule the recording.

Going Mobile with Recorded TV

If you've plunked down the money for a portable media center, you can enjoy your investment by taking your favorite TV programs on the road. Use the Sync to Device menu to copy recorded TV to your portable device.

1. In the MCE menu, select More Programs and then choose Sync to Device to open the Sync Progress screen for your portable media center.

Figure 4.12
MSN lists television
programs playing in
most United States and
Canadian communities.

Figure 4.13
Select Record This Show
to schedule your remote
recording.

Note

Your portable media center will include instructions for installing and activating the Sync to Device feature applicable to your device.

2. While the sync is in progress you can select the Edit Lists button to open the Manage Lists page.
3. Choose Add More to find the option for Recorded TV.
4. Select your recorded programs and select Save to save your new sync settings.
5. Select Start Sync to copy the program to your portable media center.

For More Information

Windows XP Media Center Edition has more than I can possibly cover in one chapter. If you need more information, there are a few books out that can fill in the gaps. You'll find them on Amazon.com, or other online bookstores. I'd be glad to write one for you as well; just let my publisher know (hint, hint).

Also, since this system is designed to be simple to use, many folks discover all they need to know just by fiddling with the remote to see what the various menus do.

COMING NEXT

In the next chapter I'll drill into one feature of Windows XP (MCE and otherwise). Windows Media Player 10 adds exciting new functionality to Windows Media management. It is available on all versions of Windows XP, either on new systems or by upgrading older systems. Get it and try it out. You'll be glad you did.

the gadget geek's guide to

Portable Media Devices

5

Windows Media Player 10

Windows Media Player is the central focus of the Windows digital media experience. It has been poked, prodded, extended, and enhanced to serve a wide range of purposes. With the appropriate skin, it serves as the Party Mode component of Windows Media Center Edition. With other trappings and trimmings, it is used in multimedia kiosks, web pages, slide presentations, and computer-based training.

In this chapter I will present the basic features of Windows Media Player and show you how to enjoy digital media at your computer. I will also show you how to use Windows Media Player to put music on your portable media player.

THE WINDOWS MEDIA SWISS ARMY KNIFE

What I'm Listening To

Title: Incantation
Artist: Delerium
Source: MSN Radio (via Windows Media
        Player 10)

Windows Media Player—or Media Player as it was originally called—has grown from a simple tool for playing back digital sound and video files to a full-fledged media management system. Not even Apple's iTunes has the power and flexibility of Windows Media Player 10 when it comes to managing audio and video libraries. With features such as CD ripping and burning (definitions later), Internet radio, music and video shopping, and library management, it has become—for want of a better analogy—a Swiss Army knife for Windows digital media.

In this chapter I will show you how to use the main functions of Windows Media Player 10. In successive chapters we will deal directly with some of its more advanced functions, as well as other Windows applications that can perform some of the same functions.

Windows Media Player 10 Features

Windows Media Player 10 has become such a large suite of applications that it is far-fetched to even call it an application. With features for music, video, recording, shopping, and even synchronizing portable media devices, it is truly a major part of what Windows XP can offer users of digital media.

Internet Radio Tuner

Even though Windows Media Player wasn't the first Internet radio tuner to find widespread acceptance, Microsoft, in their inimitable way, stuck to their guns, eventually producing what is now the foremost format for Internet radio. From the MSN Radio portal (see Figure 5.1) to Internet feeds produced by terrestrial radio stations, the WMA music format has overtaken RealAudio as a major player on the market for Internet radio content. Those stations still using RealAudio often offer WMA feeds for users of Media Player. It is installed by default on most Windows XP computers and has versions for Apple's OS/X (Mac) and Windows Mobile (pocket PCs).

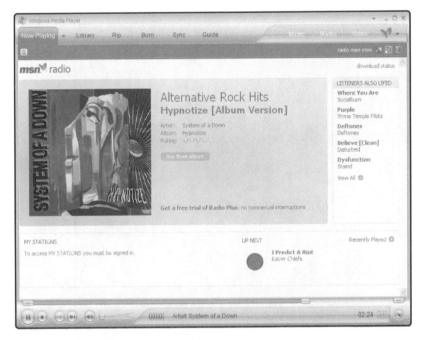

Figure 5.1
The MSN Radio portal offers Internet radio stations in a variety of genres.

Note

While RealAudio is by no means a dead format, there is not a widely used mobile component to this format. For this reason I will not spend a lot of time covering this format. For more information on the features of RealAudio and RealVideo, see the Real.com website.

Internet Video

True geeks have sought digital video as long as there have been computer graphics. From the humble start of Word Perfect and DL animations (the silent movies of our time) to the AVI and WMV formats, there has been no end to the fascination we geeks have had with making things move on the screen. Now that true HD formats are being used, the general public finally sees some point to all this. Meanwhile, wives and girlfriends went neglected, dogs went unfed, plants went unwatered, all while we crouched over a screen pasting successive images into a GIF editor to make a three-second animation loop that we could display on our web page.

Modern day (read less than five-year-old) video offers fans the ability to watch streaming (live or recorded) videos, download video clips and movies for watching offline, even synch video to a portable device like an iPod or a Portable Media Center device (see Figure 5.2).

Figure 5.2
The Samsung YH-999
Portable Media Center.

MSN Video and other Internet video sites such as iFilm.com embed Windows Media Video into their web pages. This is sometimes done by inserting links to streaming video feeds or sometimes by inserting a link to the actual video file. The streaming tags direct Windows Media to contact a server that is transmitting streaming video broadcasts; the embedded video files are simply downloaded and played (see Figure 5.3).

Smart Jukebox Media Management
Windows Media Player 10 includes a new feature called Smart Jukebox Media Management. This feature allows you to automate many of the management processes that go with maintaining a music and video library. Smart Jukebox monitors folders that you specify (see Figure 5.4) to add any relevant new media content to the music library.

Figure 5.3
The Crazy Frog video playing embedded (top) and downloaded (bottom).

Figure 5.4

Setting Monitor Folders in Windows Media Player 10.

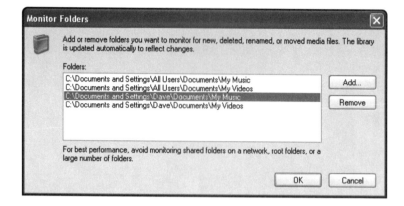

Smart Jukebox also maintains a collection of auto playlists (see Figure 5.5). In addition to the default playlists you can add your own to have Windows Media Player automatically monitor music and video additions to build playlists on the fly. Use criteria like star ratings, last time played, most/least played, and others. These playlists can then be used to build sync lists for your portable media devices or burn lists for CDs.

Figure 5.5

Auto playlists can simplify the creation of playlists in Windows Media Player 10.

CD Ripping and Burning

Windows Media Player 10 includes built-in capabilities to copy (rip) music from CD and to record music to CD for playback in compatible CD players. New in this version of Windows Media Player is the ability to rip music into MP3 format. Music in this format is directly compatible with most portable music players including the Apple iPod family of devices.

Music burned to audio CD can be played in most CD players. It is converted to a standard audio CD format before burning. Other CD burning options are available to allow the creation of Data CDs and HighMat CDs that can play in CD players designed for MP3 or WMA audio file formats.

Skins and Visualizations

If you grow weary of the whole Windows Media Player look, use a skin to make it look like a blob of goo clinging to your computer screen or a fuzzy pink—Thing (see Figure 5.6).

Figure 5.6
A collection of Windows Media Player skins.

Another type of visual element designed to mesmerize the mind while you listen to your music is the Windows Media Player visualizations (see Figure 5.7). With names like Alchemy and Plenoptic, visualizations use complex mathematics and graphics wizardry to daze and confuse.

Mobile Versions

Windows Media Player 10 includes versions for use on Mobile Windows devices and Windows Smart Phones. Pocket PCs from HP and Dell, phones from Audiovox and Samsung, and even a Palm Treo handheld communicator all offer Windows Media Player 10 Mobile.

Figure 5.7
A Windows Media Player
10 Alchemy visualization.

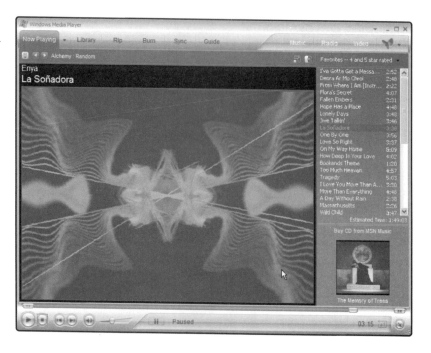

Windows Media Partners

Microsoft has built a large base of Windows Media Partners for production, distribution, and playback of Windows Media audio and video content. With their PlaysForSure program, they certify the compatible devices and merchants will work—here's a geek word for you—seamlessly with each other. This means that you can buy, download, and synchronize Windows Media content to your portable media device.

Music Stores

PlaysForSure music stores are certified to provide the ability to purchase music using Windows Media Player's built-in music search functions. A small drop-down list in Windows Media Player (see Figure 5.8) allows you to select your favorite music store as the default purchase source for your music. After selecting your music store, the Music, Radio, and Video buttons will direct you to the appropriate sections of your music provider's store. When you are browsing music selections and want to purchase media, the transaction will be handled within Windows Media Player. The media will then download directly into your music library and will be available to synchronize to your portable media devices.

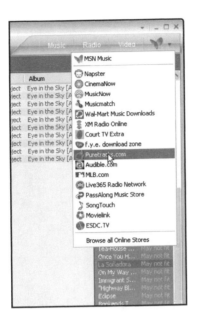

Figure 5.8
Selecting your default music store in Windows Media Player.

Device Manufacturers

PlaysForSure media devices are devices that meet the requirements for synchronization with Windows Media Player 10. You can use the PlaysForSure.com website to search for compatible devices (see Figure 5.9).

Figure 5.9
The author's Creative Nomad MuVo NX 256 is a PlaysForSure-compatible device.

Content Producers

Microsoft lists Windows Media Producer Partners on their Windows Media website. These audio and video production forms are capable of producing quality Windows Media audio and video projects.

USING WINDOWS MEDIA PLAYER 10 (THE BASICS)

What I'm Listening To

Title: Yellow
Artist: Coldplay
Source: Live365.com Internet Radio (via
      Windows Media Player 10)

It's time for an overview of Windows Media Player 10 operation. I will just cover the basics in this chapter. I'll elaborate on the most important functions of Windows Media Player 10 in coming chapters.

Playing CDs

When you insert a music CD into your computer Windows Media Player will usually open and begin playing the disc automatically. You may also choose to copy the disc using the rip feature.

Ripping CDs

To rip an audio CD, insert the disc and select the Rip tab in Windows Media Player (see Figure 5.10). Select one or more tracks to rip and click Rip Music. Windows Media Player will rip the tracks and place them in your music library. By default, Windows Media Player will also retrieve album art and additional track information from the WindowsMedia.com album information database.

Burning CDs

To burn an audio CD, insert a blank CD-R or CD-RW disc and use the controls on the Burn tab (see Figure 5.11). You can add a playlist or selections from several playlists to the Burn list. Once you are satisfied with your selections, just click Start Burn to begin the CD burning process.

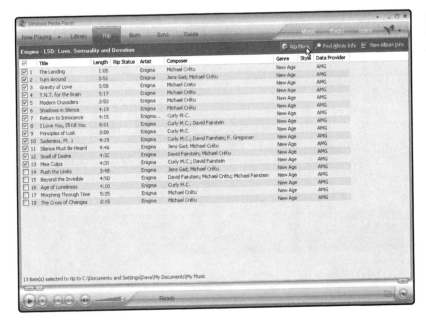

Figure 5.10
You can select just a few tracks or all tracks when ripping your audio CDs.

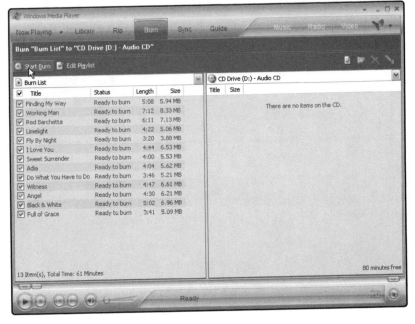

Figure 5.11
Windows Media Player 10 can burn audio CDs with just a few clicks.

Note

I will cover ripping and burning CDs in more detail in Chapter 8.

Just Playing CDs

Your CD will begin playing automatically when you insert it. If you want to adjust volume or select other tracks to play, just use the player controls (see Figure 5.12). These controls use the common conventions for Play, Stop, Pause, Next Track, Previous Track, etc. When using skins, the location and number of controls will vary.

Figure 5.12
CD playback controls in Windows Media Player.

Buying Music

Windows Media Player 10 interfaces with several online music stores. Using the stores' integrated purchasing functions you can buy and download music directly into your Windows Media Player library.

Selecting Your Music Store

Selecting your music store is a simple operation. Just click the small arrow in the upper-right portion of the Windows Media Player window (see Figure 5.13). Click on any of the available stores.

Figure 5.13
Choose your music store with this small drop arrow.

Making Your Purchase

Each store differs slightly in the way you browse and purchase music. MSN Music, for instance, allows you to purchase music using a Microsoft Passport account. Other stores have their own checkout function. The first time you make a purchase you'll create an account and provide credit card information. On subsequent visits, most stores will allow you to enable one-click ordering to make the shopping experience easier.

Once music is purchased, it will be queued for download into Windows Media Player. Folders will be automatically added to your music library, and songs, as well as cover art, will be downloaded into them. Once the download is complete, you will find your new titles in your library and in any auto playlists you might have configured that match attributes of the downloaded songs.

Listening to Internet Radio

Internet radio is free and flexible. You can choose from dozens of genres, hearing songs you'd never hear on the cookie cutter format stations littering the airwaves. You'll hear music for all tastes and in all languages. Windows Media Player has several radio portals built into the Choose Online Store list. Some, like Live365.com, offer free ad premium listening while others such as XM Radio online are subscription only.

MSN Radio

The default radio choice in Windows Media Player (of course) is MSN Radio. You can quickly choose other radio portals by making a selection from the Choose Online Store drop-down list (see Figure 5.8). MSN Radio has two options: a free option and MSN Radio Plus. MSN Radio Plus offers higher fidelity and no advertising as an incentive to subscribe.

MSN Radio organizes stations into genres, and even offers a feature where you can select stations similar to stations in your local FM radio market. This allows you to signal to MSN Radio your musical taste and receive station choices that fit that taste, but that play less format-driven selections.

Note

You are more likely to discover songs you have never heard (old and new) on Internet radio.

Other Free Radio

Other built-in choices such as Live365.com (see Figure 5.14), MLB.com (Major League Baseball Game Day Radio), and Puretracks.com offer both free and subscription options. Many stations offer the ability to click a Buy link while a song is playing to add it to your personal collection.

Figure 5.14
Live365.com Internet Radio offers both free and premium services.

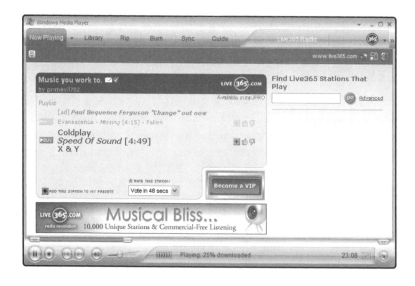

Radio Station Internet Feeds

If you do have a favorite AM or FM radio station, or know of one in another town you wish you could pick up locally, check their website to see if they have an Internet broadcast link. These links will usually launch an embedded Windows Media Player (see Figure 5.15) to play the radio station's live programming. It is a great way to bring in a favorite station when you are far from home.

Figure 5.15
A Minneapolis, MN radio station played on a laptop computer in a hotel in San Jose, CA.

Watching Internet Video

With the explosion of cable and DSL Internet connections, Internet video is finally a viable entertainment medium. Just a few years ago, Internet video was spoken of in the same breathless tones that *Popular Science* has reserved for air cars and self-cleaning windows (always just a few years away). Through the efforts of Internet access providers and the mass adoption of Internet technology the cost of broadband Internet connections has dropped to the point that many households are able to subscribe. With broadband in place, the possibility of Internet video becomes a reality.

MSN Video

As it is with MSN Radio, MSN Video is the default selection in Windows Media Player. By choosing other music stores, you can access other sources of Internet video. MSN Video offers a collection of free music videos organized by song, genre, and artist. Look for your favorite music videos or browse other selections by your favorite artists. Selecting Watch Video will open the video in Windows Media Player. After watching a short advertisement, you'll see your video. If you stay tuned after the video completes, you'll see other selections in the genre of the section you chose. You can also locate movie trailers, sports video, and entertainment videos that include selections from Animal Planet and the Discovery Channel.

Other Windows Media Video Sources

The other Windows Media Video stores offer downloadable movies, subscriptions to online movie rentals, and subscriptions to streaming video services. A quick Google search for video and the subject of your choice will often yield hundreds of movie clips that you can browse. Some will play in web pages, some are downloadable, and others may be links to streaming video sources.

Syncing Your Media to a Portable Media Device

I've covered much of the basic function of Windows Media Player. The remaining basic operation, and one reason you may have bought this book, is the ability of Media Player to synchronize media with portable media players.

> **Note**
>
> If you have a PlaysForSure-compatible video player, such as a Portable Media Center or other supported video device, you can use this procedure to sync video to your device as well.

Once your device is connected and recognized by Windows Media Player, you can use the Sync tab (see Figure 5.16) to manage the synchronization of media to the device. On the left side of the tab is music eligible for copying to the device; the right side shows music already on your device.

Figure 5.16

Preparing to sync music to a Creative MuVo NX 256 media player.

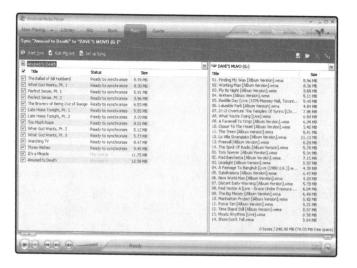

When you have selected all the songs you want to sync, click the Start Sync button to begin the sync process. Media Player will display sync progress (see Figure 5.17).

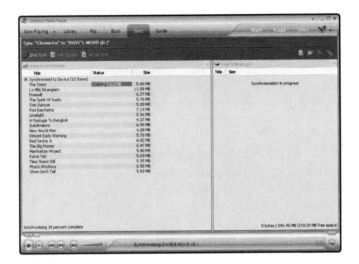

Figure 5.17
Watching a sync in progress.

NEXT ON THE PLAYLIST

While we skimmed over the process of buying digital media in this chapter, I'll give it much heavier coverage in the next chapter. We'll explore iTunes, Napster, f.y.e., MSN Music, and more. I'll show you how to protect your purchased music and deal with music licenses. I'll cover various digital rights management (DRM) schemes and help you navigate your way to the right one for you and your portable media device. In subsequent chapters I will show you how to manage your growing music library and even convert your media to play on different devices.

the gadget geek's guide to
Portable Media Devices

6

Buying Digital Media

If you're like me, you've been to some of the online music stores and had to resist the temptation to load up with tons of songs that you haven't heard for a while. Any song you care to name is available online in some form. From the Apple iTunes music service to indie artist haven Zebox, online music is a burgeoning business that is attracting a lot of attention.

In this chapter I will hit the highpoints and paddle the backwaters of the online music business. I'll show you the big music stores, revealing which are compatible with your portable media player. I'll tell you where to look for tracks you'll not find anywhere else. You'll see where the indie artists go to get noticed and where the big stars sign exclusive deals for online distribution.

ONLINE MEDIA STORES

> ### What I'm Listening To
>
> Title: Word Up
> Artist: Cameo
> Source: Windows Media Player 10

If you've been to the iTunes music store, you are no doubt aware that there are over two million music titles available to download to your iPod portable music player. If you own an iPod, this is your music store. The iTunes media player application can also play your music on Windows and Mac PCs.

If you have another brand of portable media device, you'll be using another music store. Several major online music stores cater to the non-iPod crowd: Napster and f.y.e. with over one million tunes each, Musicmatch and MSN Music each around one million, and even retail behemoth Walmart.com with around 400,000 tunes at 88 cents each.

iTunes

iTunes may be the only game in town for iPod users, but oh, what a game it is! With over two million titles, there is a good chance you'll find what you're looking for here. Each song is 99 cents, albums are typically $9.99, videos for the 5G iPod have been recently added; life is good for those with iPods and plenty of download money.

iTunes uses the iPod proprietary AAC audio codec for all their songs, making them incompatible with other media players, and by the same token, the iPod only plays MP3 and Apple codecs making it incompatible with the majority of other music stores. (If you are able to locate MP3s or non-protected WMA files, you can convert these to the iPod AAC format using iTunes.)

iTunes (the music store) is accessed through the iTunes media player application (see Figure 6.1), which is installed on your Windows or Mac computer. iTunes integrates the music purchase and download experience to ease the acquisition of music files. If your iPod is connected during the purchase, the new songs can even be automatically synchronized to your iPod, adding them to the applicable menus and auto playlists.

Figure 6.1

iTunes and the iPod are Apple's one-two musical punch.

Napster

Napster is another downloadable music browser. Using the Napster browser you can browse, search, and buy music from the Napster service's over 1.4 million title collection. Songs downloaded from Napster can be used on Windows, Mac, and Windows Mobile computers; downloaded to portable media devices; and played through Windows Media Extenders using Windows XP Media Center Edition 2005.

Napster was notorious in the late 1990s and early this century as the prime hangout for music pirates sharing their massive media libraries. Finding court-ordered legitimacy in 2003, Napster has used its notoriety to good advantage, rocketing back to the forefront of the online music business.

Napster Full Version

Napster is a Windows-only music application that interfaces with the Napster online music service. The full version of Napster (see Figure 6.2) works in much the same way as iTunes: downloading music into its music library but syncing with WMA-compatible media players.

Napster uses Microsoft's Windows Media digital rights management (Windows Media DRM) to copy protect music, but allows use on up to ten computers.

Note

The full version of Napster is primarily a subscription music service, offering users full access to over one million songs for one monthly fee. I will discuss this arrangement later in this chapter. Napster Light is the standard 99 cent per song music store like iTunes and MSN Music. In fact, Napster light is the version of Napster you'll be using if you browse the Napster service from within Windows Media Player.

Figure 6.2

Napster has gone legitimate with a full-featured music player.

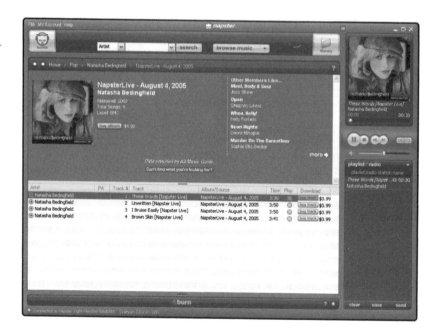

The Napster full version also includes a component for Windows Media Center Edition (MCE) that enables users of MCE to browse and purchase music using their Windows Media remote control (see Figure 6.3).

Napster for Windows Media Player

The Napster plug-in (see Figure 6.4) works inside Windows Media Player to let you browse and download tracks directly into your music library. This plug-in uses the Napster Light service to download 99 cent tunes directly into Windows Media Player.

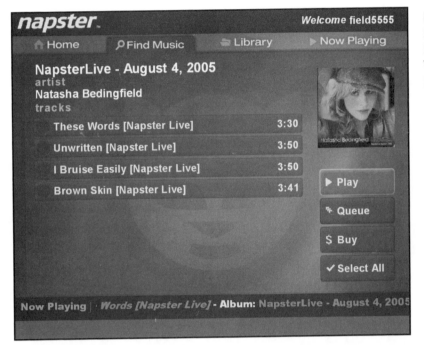

Figure 6.3
Napster Media Room Edition enhances the Windows Media Center Edition music experience.

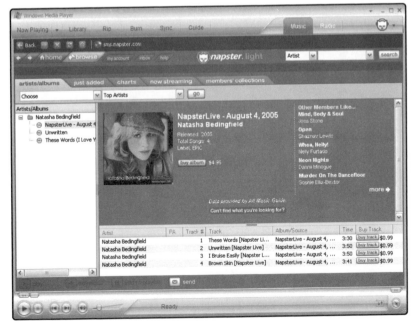

Figure 6.4
Browsing music with the Napster Windows Media Player plug-in.

In addition to the ability to search, browse, and buy music, the Napster plug-in interfaces with Windows Media's Now Playing tab (see Figure 6.5) to offer additional information about the current artist and to offer suggestions for additional artists that you might enjoy.

Figure 6.5

Napster is unfailingly helpful—offering additional information about current artists and tips on other artists whose work you might enjoy.

If that wasn't quite enough, Napster also offers Napster Radio via the Windows Media Player Radio tab (see Figure 6.6). With over 50 stations available, and the ability to buy any currently playing track, Napster Radio links entertainment and commerce very effectively.

MSN Music

MSN Music (see Figure 6.7) is the default music store selection for Windows Media Player (of course). With over one million available titles, an extensive video collection, and choices of free and premium radio, MSN Music offers a complete range of digital media choices.

Accessed primarily from within Windows Media Player, MSN Music allows you to browse by genre or search by artist, album, or song. You can purchase music from within the MSN Radio section, allowing you to immediately acquire songs you like (see Figure 6.8). As each title is downloaded, it is stored in the Windows Media library and is available to sync to your portable media device.

Figure 6.6
Napster Radio offers over 50 stations of commercial-free music.

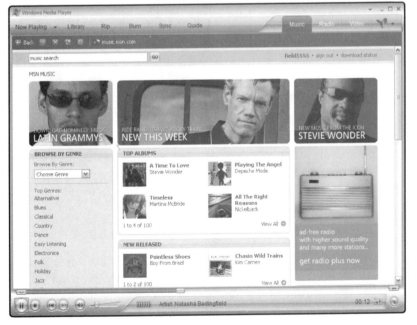

Figure 6.7
MSN Music offers over one million songs and over four thousand radio stations.

Figure 6.8

Selecting the Buy Music Playing button on the BostonPete.com MSN Radio station takes you to the MSN Music page for the artist currently playing.

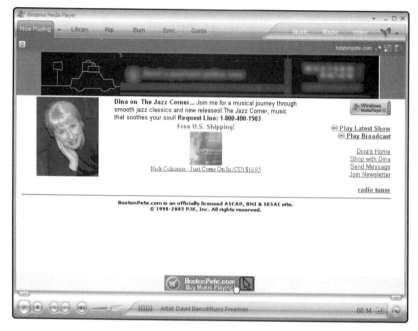

MSN Music Assistant

The workhorse that powers MSN Music is the MSN Music Assistant. This Windows Media Player extension enables the Buy features within MSN Music. It is provided separately to prevent the perception that Microsoft is forcing Windows users to use the Microsoft-owned MSN Music store.

When you attempt to purchase your first MSN Music title, the MSN Music Assistant will be installed on your system. Windows Media Player will launch the MSN Music Assistant installation page. MSN Music Assistant installs as an ActiveX Control (see Figure 6.9).

During installation you might be prompted to run the install program. The installation proceeds automatically after that. After installation, you can purchase music by clicking the Buy button next to each song's title. The first time you make a purchase, you will enter your credit card information. On subsequent purchases, the Buy button will merely change to a green Confirm button. Clicking the Confirm button completes the purchase.

MSN Radio and MSN Radio Plus

With nearly four thousand free stations and over two hundred premium stations, MSN Radio and MSN Radio Plus (see Figure 6.10) bring together an amazing array of musical preferences. From fan-favorite stations dedicated to music appreciated by fans of a single artist to stations with decade-based or genre-based playlists, there is something here for everyone.

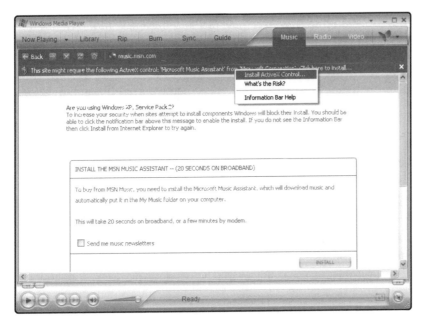

Figure 6.9
Click Install ActiveX Control to launch the MSN Music Assistant installer.

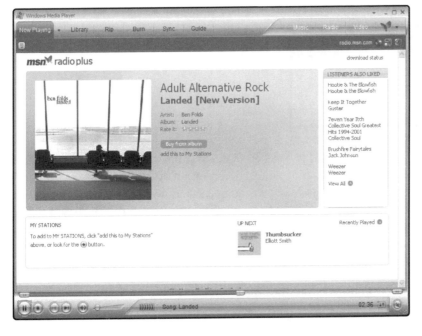

Figure 6.10
MSN Radio Plus offers high-quality 64Kbps audio streams for a monthly or yearly subscription.

MSN Radio is free and built into the MSN Music service offering. MSN Radio Plus is a subscription-based offering costing about $5 per month, or $30 per year. The principle difference is the bit rate of the music transmitted over the station's live feed. The MSN Radio Plus stations use 64Kbps Windows Media Audio, while the free stations vary between 16Kbps and 29Kbps. Both services give listeners the choice to buy the music being played through the MSN Music online music store.

Other Online Music Stores

I don't mean to concentrate too exclusively on two choices among scores—even hundreds—of online music shops, but I cannot possibly do them all justice in these pages. I picked on Napster and MSN Music for their experience and compatibility, but there are several on the PlaysForSure certification list and many others that do a great job of selling and downloading music.

Those sites that have not attained PlaysForSure certification (and that are not iTunes) may require a bit more work to download and sync music to a portable device, but may offer better prices and different selections. Zebox, for instance, specializes in edgy new music from independent artists. Many of these artists will not be seen on iTunes or MSN Music yet, though both of these mainstream sites do also encourage these artists to list their tracks.

SUBSCRIPTION MEDIA SERVICES

What I'm Listening To

Title: Every Breath You Take
Artist: The Police
Source: Nullsoft Winamp 5.11

If you're wondering how people can afford to load 30,000 songs into an iPod at about a dollar each, you're not alone. Most music aficionados would much rather spend the $30,000 on audiophile hardware than a collection of songs. What would you say, however, to having the ability to fill your player for less than $200 per year? Napster, f.y.e., and other subscription-based stores offer you access to their entire music catalog, all you can eat, for one monthly fee. For the cost of 200 songs you have access to millions!

Movie sites like CinemaNow offer digital movies as well, expanding the possibilities of the subscription model.

CinemaNow

Movies online! CinemaNow (see Figure 6.11) offers movies for rent, purchase, or by subscription from their library of over 4,500 titles. For about $4 per title you can rent movies to view on your computer. These titles work for up to 48 hours before being deactivated. Alternatively, for about $30 per month, you can download and view unlimited titles.

Figure 6.11
The CinemaNow video store is accessible from within Windows Media Player 10.

Portable Media Movies

CinemaNow also offers titles encoded specifically for playback on Portable Media Centers and smart phones. These titles are optimized for lower bandwidth playback on a portable screen. The smart phone titles are primarily music videos, but there are some sports and entertainment productions in this size.

High Definition Video

On the opposite end of the spectrum are the HD videos (see Figure 6.12). These videos are produced for Windows Media High Definition and use Windows Media Player exclusively for playback.

Napster

But wait! Wasn't I just talking about Napster? Well, yes I was. As it happens, Napster also offers a subscription version of their service. For about $15 per month, you can download any of

Figure 6.12
Windows Media HD playing in Windows Media Player 10.

Napster's over one million titles and play it on your portable media player. For most of us this is the only way we'll ever get the thousands of songs our players are supposed to hold.

Napster also offers a one-year subscription for those with no fear of commitment. Just don't let your significant other know; you might give him/her ideas!

f.y.e.

f.y.e. (see Figure 6.13) stands for "for your entertainment." Another PlaysForSure online music store, f.y.e. offers purchase and subscription options similar to those provided by Napster, with the additional option of a three-month term for those who aren't sure about a whole year.

DIGITAL RIGHTS MANAGEMENT

What I'm Listening To

Title: Breathe (2 AM)
Artist: Anna Nalick
Source: MSN Radio Plus

Figure 6.13
f.y.e. offers monthly, three-month, and annual subscriptions in addition to direct purchase.

The late 1990s and early 2000s were times of rampant music sharing. Record companies watched helplessly as thousands of their recordings were shared through the use of file sharing tools such as Napster, Grokster, and KaZaA. Legal fights led to the banning or modification of the principle file sharing programs, but the battle rages today in other mediums. BitTorrent, a nearly untraceable file sharing protocol, still shares gigabytes of files every day.

Eager to outpace the pirates, music companies have enlisted the help of computer scientists to create digital "keys" to lock their music. Authorized users get access to the media, but those who have acquired it illegitimately do not receive the key and cannot use it. This digital rights management (DRM) is the best hope for music companies eager to restart the revenue streams lost during the file sharing years.

Which DRM?

Not surprisingly, there have been various flavors of DRM produced by different companies. Microsoft's Windows Media DRM and Apple's FairPlay DRM are the leading DRM standards in use today. Most music purchased online uses one or the other DRM flavor. Not that the pirates are giving up. As you might expect, some are working to offer tools to remove your DRM protection, allowing you to share your files again. Also, some privacy advocates see the DRM as a denial of our basic freedom to use and enjoy our digital media in any way we see fit.

Apple FairPlay DRM (iTunes)

The Apple FairPlay DRM encrypts the audio channel in the AAC media files downloaded from the iTunes music service. The encrypted file can only be played by an authorized copy of iTunes or an iPod that has been loaded by an authorized copy of iTunes. A key is stored in each media file that can be used to decrypt it, but part of the key remains with Apple and is only accessed when the file is to be played or copied to an iPod. The file is then unlocked and is accessible only on that device.

Microsoft Windows Media DRM

Microsoft's DRM also encrypts the audio file. Users make use of a key granted by Microsoft when the media is purchased and which only works on the authorized systems. When an encrypted file is copied to a portable media device, the device also receives a portable copy of the key.

Other DRM Initiatives

RealNetworks's DRM (code named Helix) is probably the next best-known DRM scheme. It has not caught on and is currently only supported by the Creative Zen Xtra portable media player. RealNetworks also created a means to convert Helix-encoded songs to Apple's FairPlay DRM, but was thwarted by Apple when they made a modification to iTunes to prevent converted files from playing.

Using DRM Protected Media in a Portable Device

DRM makes portable media a bit more challenging because you can no longer just copy a media file and go. You must use an application like iTunes or Windows Media Player to copy the protected songs to ensure the key is transferred as well. You also need to make sure your portable player supports the DRM version you are using.

Locating Compatible Devices

Devices compatible with Apple's FairPlay DRM are, as you might guess, all made by Apple. The iPod family of devices supports the necessary programming to store and use the FairPlay key. Microsoft's Windows Media DRM is licensed to device manufacturers and is present on any PlaysForSure certified media player.

Dealing with Incompatible DRM Formats

Want to use your tunes from iTunes on your PlaysForSure device? The legal way to accomplish this would be to burn your tunes on a CD and rip them in your other application. There will be some inevitable loss of quality, but you will be able to play your songs on other devices. There are some tools currently available that allow the removal of DRM, but their use is controversial and not likely to be smiled upon by the creator of your DRM.

Protecting Your Investment in Media

Well, you've built a sizeable media library. You enjoy listening to it of an evening, carrying some songs on your morning run, and even gazing at the album art when you should be writing your latest book.

What would happen to your music if, suppose, lightning struck the light pole across the street while you were gazing at said album art? If your hard drive crashed you might have a lot of music stored on a disk you can no longer access. If your motherboard fried, you might see the music files on your new system but not be able to open them due to a lost DRM key.

Backing Up Your Media

If you back up your media files to CD or an external backup disk, you are taking the best step you can to ensuring they stay with you. Without the actual media files, no amount of DRM management will get them to play. They're just not there; end of story.

To back up your media you just need to copy the files from your system to a backup device. This can be an external hard drive, a tape device, or even a CD. This will ensure they are available to restore. DRM can be dealt with separately as long as the files are saved.

Backing Up DRM

With Windows Media Player you can back up your DRM license keys. On the Tools menu, choose Manage Licenses. You'll see the Manage Licenses window (see Figure 6.14) where you can select a backup location and perform a backup or restore. If you have not backed up your DRM keys, do not fear. You can often recover them from the site where the songs were purchased.

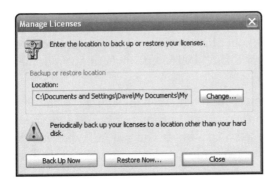

Figure 6.14

You can back up your Windows Media DRM keys using Manage Licenses in Media Player.

Apple iTunes does not back up your DRM keys. iTunes is "authorized" to play songs when you sign up for the iTunes service. You can also authorize up to four more systems to play downloaded content. To restore a system, you simply authorize it after reinstalling iTunes.

Copying Your Media to a New System

If you are sick of your current system and want to move everything to another computer, you can use the procedures detailed above to back up and restore your media and DRM. If you are using iTunes, authorize your new system and deauthorize the old one. This will ensure Apple is aware that you are no longer using the old system.

Restoring DRM Media onto Your Computer System after a Crash

If you have backed up your DRM licenses, you can restore them using the Manage Licenses tool in Windows Media or by authorizing your new system in iTunes. If you did not back up your DRM or if you have five iTunes systems authorized, you may need to contact your music store to unlock your songs.

Windows Media stores can usually provide a new DRM upon proof of purchase. MSN Music, for instance, automatically launches to do this when Windows Media does not detect a license.

If you have no available systems to authorize in iTunes, you can use the iTunes website to deauthorize all systems. You can then authorize each system you are currently using to restore full function.

LOOKING BACK (AND AHEAD)

Well, I've shown you how to find and purchase music for your portable media device. You've learned how to manage digital rights management for your music and perhaps have begun to form an opinion about how you buy and manage music in the future.

With all the difficulties surrounding digital rights management, many users simply opt to continue purchasing CDs and ripping them to disk for use on their portable media devices. If you are the owner of both Apple and non-Apple media players this may be your best path to quality digital music for both devices. If, on the other hand, you want to be a child of the 21st century and do not want a CD in the house, you have some decisions to make about your media purchases and how to protect your investment.

In the coming chapters I will give you more information about alternative music acquisition, management of your media library, and even converting music from one format to another. I fully intend to make you the best informed media buyer you can be. What you do with that information may help shape the direction major media companies take in the future to keep their increasingly sophisticated customers happy.

the gadget geek's guide to
Portable Media Devices

7

Money for Nothing and the Clips Are Free

The biggest appeal of the digital audio revolution was the sudden ability to digitize your entire CD collection and share it with your friends. Who among us who had a dual cassette player didn't make "backup" copies of their music? That was fine with the record companies, though, because the copies were never as good as the original. With digital music, however, it was suddenly possible to make exact copies of each song. Music sharing sites popped up across the Net. Napster, Grokster, and other file-sharing technologies enabled sharing on a massive scale.

Well, sharing is coming under control as better protection technologies are created. Music bought online today comes with copy protection in the form of Digital Rights Management. The purchaser can play the song only on certain registered systems, and cannot share it.

In this chapter, I will show you where the free music still lives (legally). I'll introduce you to free music directories, indie artist sites, and search engine tips you can use to locate free music. I'll show you Winamp, the grand daddy of digital music players, still going strong today in version 5. I'll also show you how to use Winamp to locate and listen to vast libraries of free music.

FREE MEDIA DIRECTORIES

What I'm Listening To

Title: Caravan to Midnight
Artist: Blue Cyberia
Source: Zebox.com
Player: Winamp 5.11 Pro

Without the impetus provided by free music and music sharing, the digital media revolution would have been more of an evolution. The desire to copy songs from CD and share them with friends (getting other songs in return) was a powerful incentive to spend long nights perfecting codecs and designing file-sharing programs.

Music sharing, as early adopters knew it, is more or less a thing of the past. On the other hand, free music is still alive and well! As I write this chapter I am listening to free songs downloaded from sources located on Red Ferret, AOL Music, Garageband.com, and Zebox. Independent artists and those hoping to break into the business publish their songs online in this forum. These are not just songs put up by rock star wannabes, however. These songs are well produced and many of the musicians are very talented people who just happen to like their day jobs. Several of these sites even have Top 10 or Top 40 lists that let you easily locate the better songs.

RedFerret.net

The Red Ferret Journal is a site specializing in technology news. You'll find articles regarding the latest digital music innovations, computer programs, or cool Google tricks. In addition to articles on interesting new technologies and devices, there is a portion of the site dedicated to free music portals and sites (see Figure 7.1). This portion of the site is maintained in wiki format, enabling users to add their own links to the project.

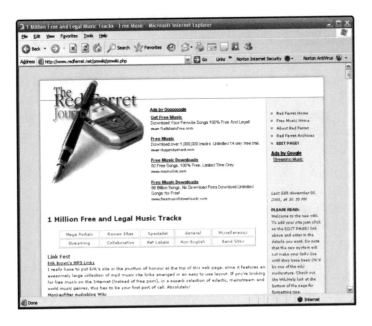

Figure 7.1

Red Ferret Journal's Free Music wiki boasts over one million free music tracks.

Wiki

A wiki is an Internet site, usually dedicated to some form of knowledge sharing that lets users make modifications to published pages. Popular examples of wikis include Wikipedia and WikiWikiWeb. The word wikiwiki comes from the Hawaiian word for fast.

One of the most informative sites devoted to wikis is, not surprisingly, a wiki itself. The WikiWikiWeb at the Portland Pattern Repository was the first wiki and can be found at http://c2.com/cgi/wiki.

Red Ferret Journal is more than just a free music site. This blog-style website covers technology news. It is not uncommon to see articles reporting cool new uses for Linux side by side with propane-powered boot dryers. Look for the "1m free music tracks" link on the site's main page.

This almost-hidden repository opens to a page of reader maintained links that will connect you to a ton of free music. You'll find links to indie artist sites, free music portals, playlist sharing sites like webjay, small record label sites, and even Spanish, Japanese, German, and French music sites.

If you happen to come across a site not listed in Red Ferret, be sure to stop back by and add your own listing. This "pay it forward" altruism is part of the spirit that has existed in the Internet since the beginning.

What Happened to MP3.com?

MP3.com began in 1998 as a music sharing site dedicated to letting music groups upload MP3 formatted audio files of their songs. Listeners could search by a number of criteria to find songs they wanted to hear. The songs were free for the downloading and many groups built a loyal following in this way. When they finally inked recording contracts, they had fans at their shows who knew all the words. Established artists used MP3.com to promote new albums; new artists found fame on the MP3.com charts; listeners got tons of free, quality music.

Things changed when MP3.com offered a new service that allowed users to upload their ripped audio files for safekeeping. The idea was that they would be able to download the audio files to their other computers and enjoy them anywhere. This feature was soon abused by music pirates as a way to share music with non-licensed people. Lawsuits were filed, MP3.com lost, and the dot com bubble burst (not all in one day, but pretty close together in geologic time). MP3.com was acquired and dismembered by Vivendi Universal and the domain name went to CNet.com, where it has returned to a somewhat useable (but no longer free) music site. Another website (Garageband.com) began offering artists the ability to advertise their songs again and life moved on.

music.download.com

A resurrection of the spirit of the popular MP3.com music site, the music.download.com music portal (see Figure 7.2) carries a large number of free music selections. Free tracks are provided by artists to allow you to sample their music before you buy it. This site includes the entire downloadable MP3 file of each song rather than the 30-second sound byte you'll find on the likes of MSN Music and iTunes.

music.download.com offers simple to use search features and the ability to browse songs by genre and geography (see Figure 7.3). Find tracks by bands you can go out and see tomorrow evening!

Figure 7.2
music.download.com is the legitimate heir of MP3.com.

Garageband.com

A recipient of some of the assets of the former MP3.com, Garageband.com (see Figure 7.4) remains a true independent artist's site. A true star-maker site, Garageband.com has an advisory board that includes such music luminaries as Sir George Martin, Brian Eno, and Steve Lillywhite. Don't get me wrong, this site is definitely in business to make money and has close ties to the music industry, but it manages to preserve much of the independent spirit of the garage band.

Search Garageband.com by artist, genre, and influence; rank songs and see if your favorite groups are moving up the charts. Top-rated artists get play on Live365.com Internet radio and exposure to major record labels. Garageband.com claims to have had 15 bands signed to major record deals. Artists such as Bo Bice and bands Tripwire and Roman Candle listed on Garageband.com before landing their own recording contracts.

Figure 7.3
Listing bands by geography lets you find acts you'll actually see in local clubs.

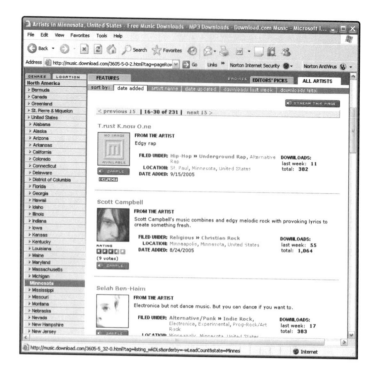

Zebox.com

For pure independent spirit, look to Zebox.com (see Figure 7.5). This is where you'll find the true-to-the-music independent artists. Bands like Project X and Third Realm live here, sharing multiple tracks with loyal listeners. Artists design their own pages, upload their own songs, and nominate their own tracks for showcase status.

Search Zebox.com by artist, city, and genre. Download and enjoy unlimited tracks. If you like an artist or band, check the gig finder to see if they will be playing in your area.

INDEPENDENT ARTISTS' SITES

What I'm Listening To

Title: Zimbabwe
Artist: Bob Marley
Source: AOL Music
Player: Winamp 5.11 Pro

Figure 7.4
Garageband.com gives many young artists their big break.

Some artists avoid the appearance of selling out and will not be found even on the indie artist sites. These artists often build their own site and put up songs for their fans. Locating a rich source of these sites is the mother lode for the independent music fan. You'll hear edgy new stuff that you'd normally only hear in back alley clubs.

Independent sites can be among the hardest sites to find. By their very nature they are elusive. Bands share their web address at gigs, fans share them online, and slowly the word gets out. As soon as traffic picks up, the band's free hosting provider pulls the plug because they're using too much bandwidth and the band has to move their songs to another site. The process repeats again and again, often frustrating the fan who moves out of town and can no longer find the band's site.

One way to locate these sites is through the facilities of Internet search engines. If you know the correct spelling of the band's name (not always an easy trick), and maybe the titles of a few songs, you can usually zero in on their site pretty fast. Use Google tricks like placing a plus sign in front of each word to get only sites using all the words in your search (see Figure 7.6).

Figure 7.5
Zebox specializes in indie artists.

Most independent artists upload preview tracks of their songs to sites like Zebox.com and Garageband.com. Search on these tracks to locate more selections the artist may have uploaded elsewhere. If the artist you are seeking is too independent to upload preview tracks, you might have to be content with what they are willing to share on their site. Even in the Internet age it is possible for artists to seek obscurity so fervently that they actually find it!

SEARCH ENGINES AND MUSIC PORTALS

What I'm Listening To

Title: The Good, The Bad, and The Ugly
Artist: Yo-Yo Ma
Source: AOL Music
Player: Winamp 5.11 Pro

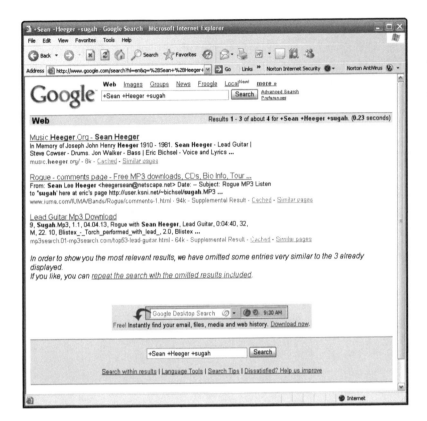

Figure 7.6
Search on artist name and a song name to find other sites and selections for that artist.

I've touched on search engines as a way to locate indie artist tracks. This is really only the surface of the treasures you can find on the Internet. With a bag of Google tricks you can unearth an amazing variety of free music. In this section I will tour the mainstream parts of the major search engines. I will also share search tricks likely to net some elusive MP3 tracks. Be sure to use extreme caution when using these tips to avoid falling victim to hackers and spyware.

Meltdown

Use caution when browsing sites returned by search engines. Many hackers put up decoy sites for free music buffs that contain nothing more than a host of viruses and spyware that attacks your system when you begin clicking their links. Other sites use these tricks to attract you to advertising pages.

Use quality anti-virus and anti-spyware programs to protect your computer and use caution when you are clicking some of these links. If something doesn't look quite right, run away!

Google Music Directory

Google Directory's Arts > Music listings are a good place to start when searching for new tracks. Try Personal Pages (see Figure 7.7) to benefit from the efforts of music pioneers who have located their own favorite artists and tracks. In these categories you'll find fan sites, artist sites, and sites established by hard-core music seekers. Learn from others who know; an empty iPod is a powerful motivator!

Figure 7.7

Google Arts > Music > Personal Pages is a good jumping off point for a music search.

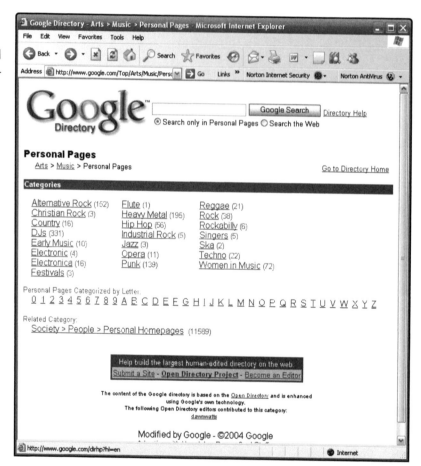

Yahoo! Music

More of a mainstream music site, Yahoo! Music even has its own music player to compete with the likes of Windows Media Player and Winamp. Don't fall for this newbie trap. You'll soon discover the music player is very good at finding more songs to buy (from Yahoo! Music, of course!).

That said, you'll find sample tracks and free music if you're patient. Also worth mentioning are the free LAUNCHcast radio stations (see Figure 7.8) and free music videos from major artists.

Figure 7.8
Yahoo! offers free commercial free LAUNCHcast radio.

AOL Music

AOL Music (music.aol.com) may be another mainstream site, but with its link to Winamp it offers some surprising free selections if you know where to look. First is the Free Music section (see Figure 7.9) found on the Songs menu on the site's home page. This section usually lists twenty or so free tracks and is changed frequently to keep folks coming back.

Even more interesting is the selection available through AOL's back door. Using the Winamp media player, you can access thousands of free music tracks via Winamp Music (see Figure 7.10). This service actually plays content directly from the AOL Music site through streaming audio. The selection is fairly stable and this is a great introduction to an amazing array of musical talent you might not experience in any other way.

Search Tips

Most search engines use specially formatted search strings to locate specific information. I'll share a few of them with you here.

Figure 7.9

AOL Music offers some free selections in its music portal.

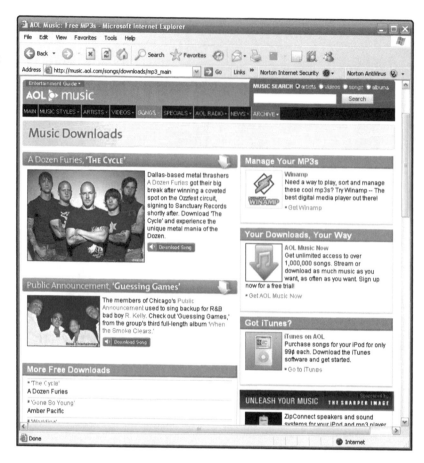

Inurl: The inurl: tag allows you to specify a string that must appear in the URL of the pages returned by the search engine. Including the file name extension can uncover a cache of music files. Even more useful sometimes is the ability to exclude certain URLs by using a minus sign in front of the inurl: tag (-inurl:htm excludes all .htm files which are the files used by normal web pages).

Site: The site: tag restricts the results of your search to a specific site (see Figure 7.11). If you want to zero in on specific MP3-rich sites, use this tag (use "site:Zebox.com MP3" to locate MP3 files on Zebox.com).

Punctuation: Various punctuation marks and other symbols can be used to tweak a query. For example, quotation marks can be used to enclose a string that you want to look for intact ("dark side of the moon"). Plus and minus signs can be used to make certain strings mandatory or excluded in the search results.

Figure 7.10

Winamp lets you access streaming music files directly from AOL Music through the Winamp Music menu.

WINAMP

What I'm Listening To

Title: Seven Deadly Sins
Artist: Flogging Molly
Source: AOL Music
Player: Winamp 5.11 Pro

If you've been paying attention to the What I'm Listening To sidebars in this chapter you've probably noticed that I am using Winamp 5.11 Professional as I write. No discussion of free music would be complete without including Winamp! One of the most respected music player applications in existence, Winamp has been the player of choice for digital music lovers since its first release in 1997. Version 2 of this player is one of the most downloaded computer programs in history (probably *the* most downloaded music application). Winamp holds the distinction of being the only mainstream music player to be able to play both the Microsoft Windows Media and Apple iTunes music formats. The professional version of the player ($19.95) can actually transcode these files to allow users to translate their Apple iPod music to Windows Media and vice versa. With the professional version, it is possible to build a unified music library containing all music purchases regardless of their source.

Figure 7.11

The site: URL restricts your search to one specific site.

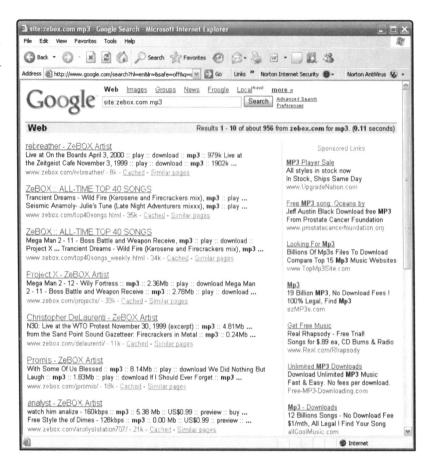

Installing Winamp

Winamp can be downloaded from a variety of sources, but the home base for this media tool is Winamp.com. Download the standard version, which includes most of the features of the professional version (see Figure 7.12), or drop the $19.95 to get the professional version immediately.

Note

If you are in a hurry to get the music playing, you can even download a "lite" version that skips most of the bells and whistles but plays all the same tracks the full versions do.

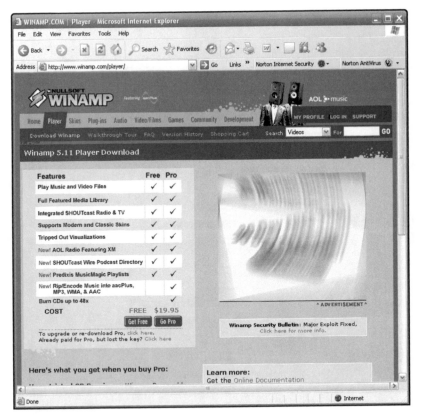

Figure 7.12
Winamp standard includes most of the features found in the professional version.

Winamp downloads an executable installer that you can run to install it on your system. Accepting the default options in the installation wizard will install Winamp with default settings that work for most systems. Winamp will swing into action immediately after installation, offering to scan your system to find media files to play. If you allow this, it will load in your iTunes and Windows Media files and add them to its music library. With the latest version and appropriate plug-ins, Winamp Pro even plays protected iTunes and Windows Media music! (There will be more on transcoding music and dealing with DRM in Chapter 9.)

Winamp Plug-ins

Winamp is extended by the use of plug-ins created by one of the members of the Winamp Developer Network. Volunteer developers have created over 20,000 skins to modify the appearance of Winamp, and hundreds of plug-ins (see Figure 7.13) that add features such as additional transcoding ability to the player. Among these plug-ins are tools for playing protected iTunes files and synchronizing with iPod portable media players, adding language features to Winamp, and even games like Pac Man.

Figure 7.13
Plug-ins can customize and extend Winamp to offer new features and languages.

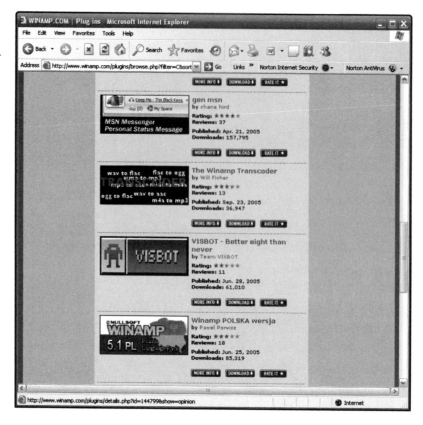

SUMMARY

Some might say free media is dead, but I think the opposite is actually closer to the point. Free music is alive and well, and thriving on the Internet. From free sample songs to free tracks from independent artists, the use of free music as a promotional tool is very much in vogue.

The use of free music directories and search engines will locate much of this free music, and link sharing by fans of independent artists is another important source. Winamp, the first major media player, is also an important hub on the use of free Internet media. With directories of free media built in and media consolidation and transcoding functions, it is a powerful media tool.

8

Ripping, Mixing, and Burning

Sounds violent, doesn't it? One might think from the title that this chapter is a guide to being a good Viking baker! It's not as bad as all that. In fact, one of the most satisfying (and nonviolent) aspects of digital music collecting is the ability to easily create playlists and mixes using your own music. If you like to create compilations of your song collection to fit your mood, create the scene for a party, or if you just want to burn a disc for the car, you'll appreciate the tools I will discuss in this chapter.

Digital music is unique in that it can be reproduced in exact detail from one medium to another. Music CDs can be copied (or ripped) with little or no loss of fidelity. The songs can be combined with other songs to create your own unique compilations. Sync the songs to your iPod or portable music player or copy (burn) them to another disc to take along on the road. Burn them in CD Audio format or simply copy a bunch of digital files directly to disc. Many car CD players can now accept CDs created with digital media players, playing the MP3 or WMA formatted songs right off the disc. This allows you to take hours of music on a single disc where the original CD format can only support 70–80 minutes.

RIPPING YOUR CD COLLECTION

> ### What I'm Listening To
>
> Title: Tom Sawyer
> Artist: Rush
> Source: CD
> Player: Windows Media Player 10

Ripping music is not the violent act that the name seems to imply. When you copy your music to your computer it is simply converted from the CD's audio format to a more compact digital format that lets you store your music more efficiently.

Some music formats, as you learned in earlier chapters, do not reproduce the full fidelity of the original tracks. There are variations in codec and bit rate that can create perceptible changes in the music. I will discuss format and bit rate selection to help you decide which is best for your needs. I will also show you how to rip music using the three most popular media players.

Media Player 10

The CD ripping features of Windows Media Player 10 offer you a large selection of available codecs and bit rates for copying your music. You can choose MP3 and WMA formats in bit rates from 48Kbps (AM radio quality) all the way up to lossless WMA which preserves every detail of the original track. Most users of Windows Media Player find the variable bit rate (VBR) settings to be the best for quality and file size. Variable bit rate encoding uses lower bit rates for slow or quiet passages, and higher bit rates when necessary to faithfully reproduce complicated parts of the music. Windows Media Player uses a quality slider to allow you to select the recording quality you wish to achieve (see Figure 8.1). You can locate the quality adjustment on the Rip Music tab in Windows Media Options.

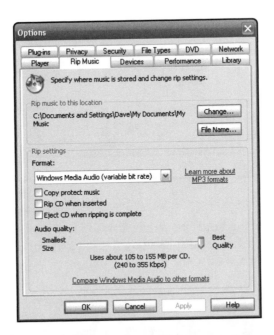

Figure 8.1

Use the quality slider in Windows Media Options to choose the encoding quality you want.

Choices, Choices

When you insert a CD into your Windows XP computer you usually receive a choice of actions that you can perform (see Figure 8.2). If you have media applications installed in addition to Windows Media Player, their options will also be displayed here. If you simply want to play the CD, you can choose to do so here. If you choose one of the rip options you will rip the disc using the default settings you have set in Windows Media Player (or the alternate applications you have chosen).

Some applications like iTunes don't use the word "rip." Instead, they use the more politically correct "import." While the thesaurus doesn't connect these two words, trust me, they mean the same thing to your computer.

Figure 8.2
Windows XP will ask you
what action to perform
on an inserted CD.

The Rip Tab

Windows Media Player 10 organizes ripping functions on the Rip tab (see Figure 8.3). When a
music CD is inserted, the tracks from the disc are displayed on this tab and can be selected for
ripping. Select the tracks you want to rip and click the Rip Music button.

Figure 8.3
The Rip tab is the control
panel for CD ripping
in Windows Media
Player 10.

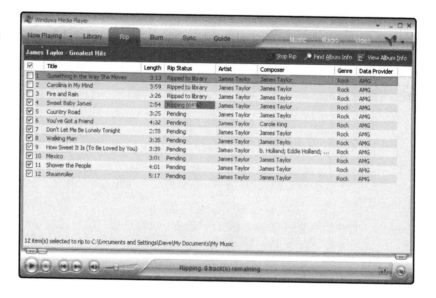

Note

To select or deselect all the available tracks, click the small checkbox at the top left portion of the window.

The Find Album Info and View Album Info buttons perform pretty much the same task, listing tracks and providing additional information about the album. The View Album Info goes one better, though. You can actually buy one or more of the tracks from the MSN music store. (Wait, don't you already have the CD?)

To Rip Songs in Windows Media Player 10:

1. Insert a CD into the computer's CD drive.

2. If the system prompts you for an action, click Cancel to close the action window.

3. Open Windows Media Player and select the Rip tab.

4. Select the tracks you want to rip and click the Rip Music button.

Note

If you simply want to rip all tracks to your music library, you can choose the Rip Music From CD option on the system action window when prompted. Windows Media Player will rip all songs using the default settings that have been configured on the Rip Music tab in the Windows Media Player Options window.

Windows Media DRM

When you rip songs from a CD, you can choose to use Windows Media DRM to copy protect them. In the Rip Music settings (see Figure 8.1) you can check the Copy protect music option to use Windows Media DRM to copy protect the songs. This will keep anyone from playing these songs on a non-licensed system. Take care when making this change, as the license keys used to encrypt these songs exist only on your system. If you ever need to move them to another computer, you will need to move the license keys as well. This could be a problem if the original system has crashed.

iTunes

As I mentioned earlier, iTunes doesn't "rip" music. Apple is much more refined than all that. With iTunes, we "import" tracks.

When a CD is inserted and iTunes is chosen to import tracks, the tracks are displayed on a selection screen. Clicking the Import CD button at the top right of the screen begins the import process (see Figure 8.4). Tracks are imported and placed in the iTunes library and added to applicable smart playlists automatically.

Figure 8.4

iTunes is a bit more genteel with its ripping process.

iTunes can import tracks in AAC, Apple Lossless Codec, AIFF (an older Macintosh format), and MP3. Tracks copied from disc do not carry copy protection and can be transcoded to MP3 or another format by tools such as Winamp to allow them to be played on devices other than the iPod. Alternatively, songs can be imported directly into a compatible format such as MP3. Default selections for import codec can be made on the Advanced tab in iTunes' Preferences (see Figure 8.5). Choose the Importing sub-tab to make these selections.

To Import Songs Using iTunes:

1. Insert a CD into the computer's CD drive.

2. If the system prompts you for an action, click Cancel to close the action window.

3. Open iTunes. Select Audio CD under the Source column heading.

4. Choose the tracks you want to import and click Import CD.

Note

The option to import all tracks is also offered as a choice when a CD is first inserted. Choosing this option will import all tracks using the current settings configured in iTunes.

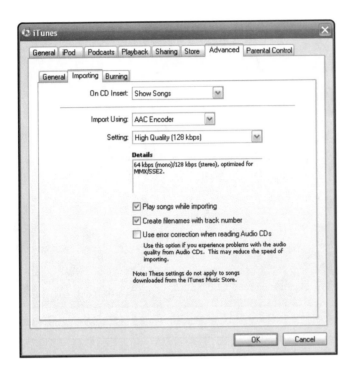

Figure 8.5
iTunes default import settings are maintained in the Preferences menu.

Winamp

As you might expect from a media player that you actually spend money for, Winamp offers a few more choices when ripping music from CD. In addition to the ability to simply rip selected tracks, the Preferences menu can be accessed from the same selection menu (see Figure 8.6). Format options are also more extensive. Default choices of MP3, WMA, and AAC can be extended by adding encoding plug-ins. These are available from Winamp's website. You can pick up encoders for OGG/Vorbis and FLAC (Free Lossless Audio Codec) there.

There are free versions of Winamp, but these come with limitations. With the free versions you cannot rip media to MP3, and ripping and burning speeds are limited. Also, without the ability to encode MP3, transcoding options are severely limited. These versions are an excellent way to see if you like Winamp, but at $19.99, the professional version removes any limitations.

When you've completed ripping tracks from your CD, Winamp creates a playlist based on the ripped tracks, and gives you an option to burn the ripped tracks to CD. This is an excellent way to make a "road CD" for the car.

Winamp offers options for playlist generation and output file generation. This allows you to configure Winamp to automatically create the appropriate tags for MP3s as they are imported to allow MP3 players to index them properly (see Figure 8.7).

Figure 8.6
Winamp lets you change rip settings on the fly.

Figure 8.7
Winamp offers extensive options for CD ripping.

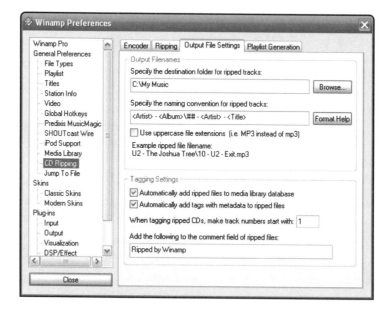

To Rip Tracks Using Winamp:

1. Insert the CD to be ripped into your system's CD drive.

2. If Windows prompts you to take an action, click Cancel to close the action window.

3. Open Winamp and select the CD drive's letter designation under Devices in the Media Library window pane. (If you do not see the Media Library, press Alt + L to open it.)

4. Hold down the Ctrl key and click the tracks you want to rip to select them.

5. Click the Rip button and choose Rip selected tracks. (If you want to rip all tracks, click Rip and select Rip all tracks from the choice menu that is presented.)

CHOOSING FORMATS AND BIT RATES FOR RIPPING

What I'm Listening To

Title: Wondrous Stories
Artist: Yes
Source: Purchased from MSN Music (WMA-
    DRM)
Player: Winamp 5.11 Pro

Choosing the encoding format and the bit rate for your music ripping is a bit like choosing the carrier for a package delivery. Some will get the package there faster, or more reliably, but you will probably have to pay more for this level of service. Some won't deliver to certain addresses, while some will deliver to any known address. Formats vary with their compatibility with various devices. Some like the Microsoft WMA format play on many players, while some play on all players (MP3). In this section I will discuss format and bit rate selection with an eye to quality and storage requirements.

Format Selection

Formats like the Windows Media and Apple iTunes formats, for all their popularity, still don't deliver to all addresses. Without changing trucks (formats), you'll never play a WMA file on your iPod. You'll have similar problems playing your iTunes (AAC) files on any Windows Media compatible device. If you absolutely need to be able to deliver your music to all known addresses, you'll want to use a format such as MP3 that can be played on both iPods and WMA-compatible players. As much as the various vendors may seem to look down upon the lowly MP3 format, they all still support it as the lingua franca of the digital music world.

An increasing number of music players also support the Vorbis format (also known as OGG/Vorbis or simply OGG). While this format improves the quality of recordings over what MP3 can achieve, it is not yet to be considered a universal format the way MP3 is. It bears watching as it becomes more widely accepted.

Sometimes you'll find yourself stuck in a particular format. Maybe you've bought a few albums on iTunes. They are stuck in the DRM-locked AAC format. There are ways you can transcode these files to any other format you wish to use. In Chapter 10 I'll show you how to accomplish this.

Note

Transcoding DRM-protected music to share with your friends is a violation of the Digital Millennium Copyright Act (DMCA) and can land you in jail. Transcoding for your own use in non-compatible devices is considered by most to be a "fair use" application for the music you have purchased and should not cause any problems. The music industry does not agree with this, but privacy advocates and music industry groups are waiting for a good "test case" to be presented to the courts to see which interpretation of the DMCA will prevail.

Bit Rate Selection

There are many religious debates regarding quality of one format in relation to another. Most fans of the AAC format will resolutely insist that there is no way WMA can even compete in terms of quality. WMAers think the opposite, obviously. Proponents of both camps scoff at MP3. OGG fights for respect from everyone. Who is right? Well, everyone is. When recorded at identical bit rates, most people would be hard pressed to tell the difference between music encoded in one format over another. The fact is, what makes the biggest difference in recording quality, and what most of these fans do not take into account when making their claims, is the bit rates of the respective recordings.

At 64Kbps, most formats will leave something to be desired. At 320Kbps or higher, only an oscilloscope will be able to truly tell the difference between two formats. Each format makes decisions based on certain psychoacoustic principles of which sound waves to leave out of the recording. Unless you have trained your ear to listen to music differently, you'll not be likely to notice which parts have been left out. As bit rate is increased, less and less of the music is left out. In fact, at the high-end, the lossless version of each format records every part of the music, producing playback that is indistinguishable from the CD original.

As you can see, the choice of format is much less important than the choice of bit rate. You'll need to experiment a bit to see where your tastes lie with respect to bit rate. You'll get a lot more music on your portable media device if you can manage to accept the quality of 64 or 96Kbps recordings over, say, 128 or 192Kbps. The choice is ultimately up to you and your ears.

MIXING YOUR MUSIC

What I'm Listening To

Title: Soft Focus
Artist: Droplet
Source: iTunes
Player: iTunes

Okay, I'm not going to make you into a mixmaster, or a DJ here. That takes skills I just cannot convey in the pages of a book. What I am going to show you how to do, however, is to create compilations of songs to suit a mood or to support a theme. Using the popular media players, you can create playlists and compilations CDs for parties, road trips, or train rides. Impress your date with a romantic mix created from songs you already own, or acquire a few tracks to add just the right atmosphere. One tool, the Predixis MusicMagic plug-in for Winamp even suggests additional titles based on a single music selection.

Creating a Mix CD in Windows Media Player 10

Windows Media Player uses playlists to organize the order in which songs will be played. By adding tracks to a playlist (see Figure 8.8) you can create your own mix. When the mix is complete, play it in Media Player, sync it to your portable media device, or burn it to a disc.

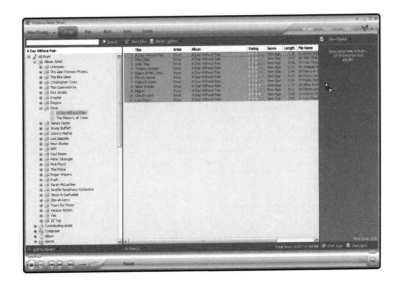

Figure 8.8
Creating a playlist allows you to organize a compilation.

The choice of songs is up to you with Windows Media Player. There are no features to offer similar tracks for your compilation. This will certainly ensure that your compilation is truly a reflection of your tastes and feelings. Be sure to remind your significant other of this as you enjoy the musical medley that you have arranged! He or she will look at you strangely for just a moment, make a quick decision about his or her compatibility with one so geeky, and (hopefully) go back to enjoying the music. If you're using your portable media device, consider bringing a headphone splitter or portable speakers. While intimate, sharing earbuds seems a little, well, cheap.

To add songs to a playlist, browse the Library tab. When you find a song you want to add, right-click it and choose Add to Burn List or Add to Sync List. This will queue it for burning to CD or syncing to your portable media player. When the selections are complete, you can use the Burn tab or the Sync tab to complete the operation.

Auto playlists (see Figure 8.9) give you the ability to create CD-sized lists of music meeting criteria that you specify. This can be music from certain artists, certain genres, even songs using certain digital media formats.

To create an auto playlist:

1. Right-click Auto Playlists and click New.

2. The New Auto Playlist configuration window will appear. Provide a name and select the appropriate criteria for your list. Add a size limit appropriate for the disc you are burning.

3. Click OK to save the new auto playlist.

Figure 8.9

Create an Auto playlist to help organize a CD's worth of music.

Mixing with iTunes

iTunes allows you to make playlists in much the same way Windows Media Player does. In addition to this ability, iTunes also can create Smart Playlists. If you configure this properly, your romance mix CD will build itself. In Figure 8.10, I have created a smart playlist that fills one 80-minute playlist with Barry White and Tony Bennett songs. When the time comes, I just burn that playlist to CD and warm up my scooter. Romance is in the air!

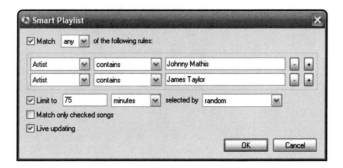

Figure 8.10
The iTunes Smart Playlist is a great way to precompile a burn list.

To Create a Smart Playlist for a Romance CD:

1. On the File menu, select New Smart Playlist.

2. Create rules by choosing criteria such as Artist or Genre.

3. Click the checkbox next to Limit to and choose a length equal to the capacity of your CD.

4. Click OK and type a name for the list.

Winamp's MusicMagic by Predixis

If you thought the smart playlist was great, just hold onto your glasses. Winamp includes a plug-in created by Predixis called MusicMagic. Its sole purpose is to create mixes based on the genre and musical stylings of a selected track. MusicMagic scans your music collection, and categorizes each track in advance. When you are ready to create a mix, simply select the track in your music library and click the MusicMagic Mix button. MusicMagic creates a quick playlist of 20 tracks (see Figure 8.11). These tracks can be played as-is or enqueued to an actual playlist. Burn or sync the playlist to take your romance on the road!

Figure 8.11
MusicMagic creates a list of 20 compatible songs based on a single selection.

BURNING DOWN THE HOUSE (MIX)

What I'm Listening To

Title: Burning Down the House (of course)
Artist: Talking Heads
Source: Ripped from CD
Player: Winamp 5.11 Pro

In many ways we have already covered much of the mystery that is burning tracks to a CD. Just a few years ago, creating a CD was a long, tedious, process. Now, most media players offer to burn songs to disc every time you turn around. Gone are the days when geeks created three coffee coasters for every disc successfully burned. I'll show you how to burn a disc with each of the three media players I've presented in this chapter. You'll see for yourself their simplicity and you'll be creating your own audio CDs before you know it.

A Word about Discs

Not all CD-Rs are created equal. When you intend to create music CDs it is important to have an understanding of the discs you need for the task.

First of all is the difference between CD-R and CD-RW technology. If you will be making a lot of temporary discs, CD-RW (CD-ReWritable) makes the most sense for you. If you like to keep your compilations forever, choose CD-R (CD-Recordable).

If you choose CD-RW, be sure to do a test burn with one disc before you go out and buy a hundred of them. Some older CD recorders have difficulty working with CD-RW formats and seem to write to disc only to turn around and show you a seemingly blank disc. If this happens to you, you might need to upgrade your recorder's firmware, get a new recorder, or fall back to CD-R discs for your music.

Speed is another factor in disc selection. It is still possible to find discs that record at only 2X or 4X (X being the speed of the first computer CD-ROM drives). Most CD burners available today can record at speeds as high as 52X. Get the fastest discs you can afford to ensure you don't have a speed mismatch. Even if you don't use the highest speed when you burn, you'll get better recordings when you use the higher quality discs. If you cannot find high-speed discs, adjust the burn rate in your media player's settings.

Burning Tracks in Windows Media Player

In true Windows fashion, Windows Media Player gives you about six ways to accomplish each task. As you read through this section, take note of the different places that the Burn function is available in Windows Media Player.

Burning from the Library

Windows Media Player offers the option to burn entire albums or selected tracks from within the music library. By right-clicking an album or selecting a group of tracks and right-clicking one of them, you can choose Add to Burn List (see Figure 8.12) to create the list of tracks you want to burn.

Once you have selected the tracks you want on the disc, click the Start Burn button to begin the burning process. When the burn is complete, the disc will be ejected and is ready to use.

Burning from the Burn Tab

The Burn tab presents a list of tracks waiting to be burned and displays the contents of the target CD. By choosing from the drop-down list in the left pane, you can pick tracks for the list (see Figure 8.13).

Figure 8.12
Burning tracks to disc can be a right-click operation in Windows Media Player.

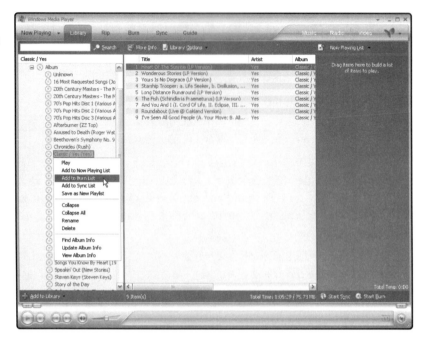

Figure 8.13
Choose tracks from the drop-down list in the Burn tab to create a burn list.

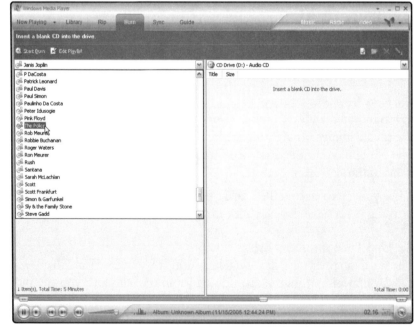

When you have selected all the songs you want to burn, click the Start Burn button to begin the burn process. When the burn is complete, the disc will be ejected and is ready to use.

Burning an Auto Playlist

One of the great features of an auto playlist is its ability to create a collection of tracks for burning all in one operation. To burn the auto playlist I created in Figure 8.9, just right-click it and choose Add to Burn List (see Figure 8.14).

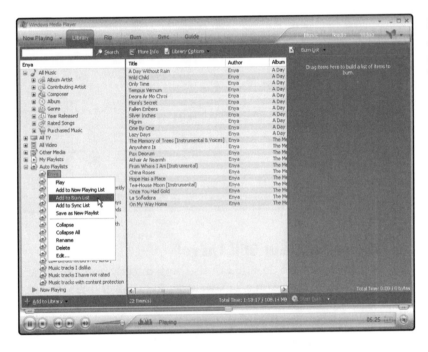

Figure 8.14
Auto playlists make it simple to create collections for burning to disc.

Complete the burn operation by clicking the Start Burn button. When the burn is complete, the disc will be ejected and is ready to use.

iTunes May Not Rip, but It Does Burn!

iTunes also has the ability to burn imported and purchased tracks to CD for use in CD players. In a way very similar to that of Windows Media Player, you can burn tracks directly from playlists. To simplify the user interface, the burn tools only work from playlists. While it may seem tedious to have to create a playlist before burning tracks from an album, the tools for creating standard and smart playlists make the process fairly quick.

When you have created a playlist and selected the songs you want to burn, click the Burn Disc button in the upper-right portion of the window (see Figure 8.15). The Burn process will begin and will proceed until all tracks are converted and burned. When the burn is complete, the disc will be visible under the Source column as a playable music source.

Figure 8.15
Once a playlist is created, the Burn process is a one-click operation.

Winamp's Burn Is Harder to Find, but Still There!

Winamp can burn from playlists as the other two do, and can also burn directly from the library. It is possibly the fastest of the three, writing a disc on a 24X recorder in just over three minutes.

Burning from the Library

To burn tracks or albums from the library, just right-click the album or track and click Send to: and choose CD Burner on D: (where D: would be the drive letter of your CD recorder). This action creates a burn list on the device, which can be accessed by clicking on the device under Devices in the library tree to the left of the Media Library pane (see Figure 8.16).

When you click Burn, a small window is displayed (see Figure 8.17) to allow you to adjust speed and BURN-proof mode of the CD recorder. Select settings appropriate to the media you are using and click Burn.

BURN-proof

BURN-proof is a technology supported by newer disc drives that helps prevent emptying the burn buffer, a section of memory in the recorder that holds data waiting to be burned. Running this buffer empty has been known to corrupt data in the disc. This condition is known as a "buffer underrun." BURN in this instance is an acronym for "Buffer UnderRuN." Ain't our acronyms great!?

Figure 8.16
Clicking Burn begins the burn process.

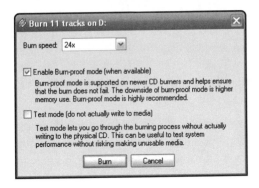

Figure 8.17
Winamp allows you to adjust burn speed and BURN-proof mode prior to beginning the burn process.

Burning from a Playlist

Clicking Burn from a playlist view (see Figure 8.18) in Winamp immediately displays the speed and BURN-proof window (see Figure 8.17). Clicking Burn begins the burn process.

Figure 8.18

Burning from a playlist is about as simple as it gets.

RIIIIP!

If you're the type to follow along on your computer, you've probably ripped a few CDs by now. Take some time to listen to your ripped tracks. Compare quality among formats. See if you can tell the difference when you adjust bit rates. Find your sweet spot.

If you take the time to do this now, you can set consistent rip settings for your growing media library. As you gain more experience with media management, you may want to rip some tracks into other formats such as Vorbis or FLAC. See how they compare with the mainstream formats.

In a few weeks it will be a safe bet that you'll have some pretty firm preferences and will be able to hold your own in any discussion of formats and codecs. You'll know what folks are saying when they discuss the relative merits of lossless vs. lossy formats. You'll impress (and slightly intimidate) your friends with this new knowledge and will probably find yourself being asked for advice in this area. You may not know it yet, but it is actually a little late to avoid the label *gadget geek*. You simply know too much at this point and from here on it's a slippery slope. Welcome to the brother(sister?)hood!

Starting in the next chapter, we will be getting into the wizardry of media management and creation. You'll enjoy the power of knowing you can control your media destiny. Even if you have a large investment in a particular media format, you can change boats midstream. You'll learn how to convert your media to new formats and how to create a consolidated media library spanning all your media players while simplifying your ability to buy and use your digital media.

9

Converting Media Formats to Play on Different Systems

Let me pose a hypothetical scenario for you.

Suppose you've got a portable media player manufactured by a large manufacturer of such devices. Suppose as well, that you've spent what amounts to the national debt of a small country on tunes from this manufacturer's online music store.

One fine day, you're out driving your Bug or your Vespa scooter and hit some railroad tracks. Your portable media buddy bounces from his (her?) secure spot on your dash (or in the case of the scooter, front basket). You make a grab for it, but self-preservation winning out, decide that gaining control of your vehicle is more important than catching your tunes. Your hand instead seemingly bats the player away like a bothersome insect, sending it leftward into oncoming traffic. The player executes a graceful arc—beautiful to the last—never hitting the ground, but imbedding itself artistically in the radiator of an oncoming tractor-trailer. The police investigate the mishap, determine you've lost enough for one day. The trucker gets a tow, grumbling something about propeller-headed something.

It strikes you as you sit morosely at your computer that evening that the only place you can play your small fortune in tunes is in your house. You spend the next weeks sitting with headphones, laughing and crying at the memory of each tender track, hearing songs you'd forgotten you had, and thoroughly depressing yourself. Your friends organize an intervention, showing you your pale skin and unkempt hair in a mirror.

"C'mon! He'd (she'd?) want you to move on!" You see the wisdom in their caring actions and decide it is time to once again join the living.

In the store you gaze at the sleek, new media players, touch a few, even stroke the face of a new video model. You realize, though, that it is just too soon. The memory is too recent. Besides, the thing costs over $400! Leaving the store, you catch sight of the nearby big box computer store. Drifting in, you see row after row of inexpensive players. It seems right (for now) to just get an inexpensive surrogate to help you through this rough patch. It's not like you're being unfaithful. After all, it's not even the same species.

You bring the new player home and connect up to your computer. You hit the sync button on your media player and wait expectantly. Nothing happens. You realize you have to load the player's favorite media player to sync media. After doing so, you try to import your songs. Still nothing; just a rude message about incorrect format. You're starting to think this is not as easy as you thought. Did your old player really expect you to suffer for eternity without him (her?). Your resolve strengthens to get this music into your new player…

The Argument for and Against Transcoding Files

Most folks would tell you that—having paid for them—your songs are yours to use as you see fit. That seems to make sense. The problem is that some proponents of the Digital Millennium Copyright Act (DMCA) see this differently. They feel that you should not be allowed to defeat copy protection on any media–even your own. They further state that what you've purchased is not the media, but the license to listen to it.

Most music sellers that sell media protected by digital rights management (DRM) allow you to unlock it to play on portable devices and on multiple computers. This still looks like a limitation to those who would have no restrictions on their rights to use the media that they have purchased. Until a good test case is brought to the courts, many media buyers continue to use their media the way they always have. They buy the CD, rip it to their computer, and play it wherever they want. They encode it with OGG/Vorbis or MP3 and, "To heck with the DRM-laden online media offerings!" As far as I can tell, this is still perfectly legal.

I cannot and will not weigh in on this dispute, but you should be informed that there are ways to be able to continue to use your DRM-protected media after the loss of your primary player, even using the media on incompatible players. This is what this chapter offers. I do not condone violating the law, and have no firm opinion—one way or the other—regarding the constitutionality of the DMCA. I simply provide this information to allow you to make your own decisions on this matter and to act accordingly.

MEDIA CONVERSION TOOLS

What I'm Listening To

Title: Sirius
Artist: Alan Parsons Project
Source: MP3 (ripped from CD)
Player: Winamp 5.11 Pro

This section deals with the features built into the major media players to allow you to convert media from one format to another. Media not protected by DRM can be freely converted using these tools. DRM-protected media requires further processing before it can be transcoded. I will cover the steps required to convert this media later in the chapter.

Windows Audio Converter

Windows Media Player has no built-in ability to transcode media files, but has hooks to the Windows Audio Converter (see Figure 9.1), a product included with Windows XP Media Center Edition 2005 and Microsoft Plus! Digital Media Edition. The Windows Audio Converter converts Windows Media Audio (WMA) format files to MP3 format and vice versa.

Figure 9.1
The Windows Audio Converter can be executed from within Windows Media Player or separately for batch conversion.

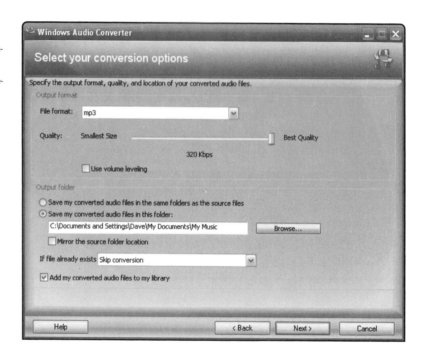

Converting Files from Windows Media Player

To convert files from within Windows Media Player:

1. Right-click the file you want to convert in the Windows Media Player library and select Send to Windows Audio Converter.

2. On the Windows Audio Converter options screen, choose the format, bit rate, and output destination for the new file. Click Next to move to the next step.

3. Click Start Conversion to begin the file conversion.

When the conversion is complete, you can play the file in Windows Media Player or sync it to your portable media player.

Note

The version of Windows Audio Converter included in Windows Media Center Edition 2005 differs visually, but has the same features as the Plus! Edition.

Converting Files in Windows Audio Converter

You can also use the Windows Audio Converter by directly executing it from the Start menu. Using this option, you can batch convert entire folders of digital media files.

To execute Windows Audio Converter directly:

1. Execute the Windows Audio Converter Start Menu shortcut.

2. On the first window, choose either the option to convert an entire folder or to convert specific files (see Figure 9.2). Click Next to continue.

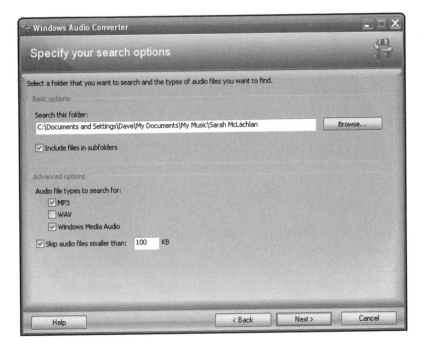

Figure 9.2
Windows Audio Converter converting a batch of files at once.

3. If you are converting a batch of files, you can remove certain files from the batch on the next screen (see Figure 9.3). Click Next to continue.

Figure 9.3
Remove any files from the conversion batch that you do not want to convert.

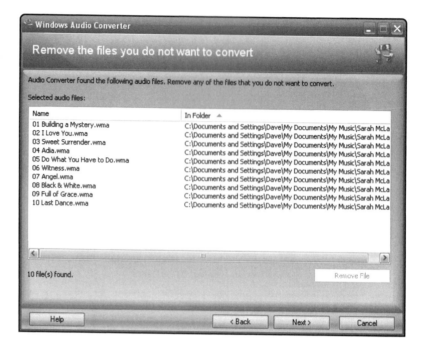

4. Choose format and bit rate options for the converted files. This option is the same as for conversions initiated from within Windows Media Player (see Figure 9.1). Click Next.

5. Click Start Conversion to begin the file conversion.

When the conversion is complete you can play your converted files in Windows Media Player or sync them to your portable media player.

iTunes

iTunes excels at converting other formats to iTunes playable formats. You can convert WMA, MP3, WAV, AIFF, and Apple Lossless formats to AAC, MP3, AIFF, WAV, and Apple Lossless.

Note

Notice the lack of WMA as an output format. iTunes can play all but the WMA format natively. WMA must be converted before playing. WMA-DRM cannot be converted. I will discuss ways to convert DRM formats later in the chapter.

In iTunes, the conversion can be done in one of two ways, depending on the format of the source files. Conversion of WMA files must be accomplished during the library import process. All other formats are imported without conversion and must be converted after the import process is complete.

Converting WMA Formatted Files

When you choose to import WMA files into iTunes, the conversion happens immediately.

To import (and convert) WMA files:

1. On the File menu, choose Add File to Library or Add Folder to Library.

2. Browse for the file or folder you want to add to the library (see Figure 9.4). Select the appropriate files or folder and click Open.

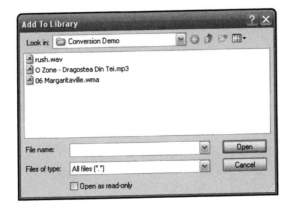

Figure 9.4

iTunes can import files in several formats, but only converts WMA during the import process.

3. iTunes will advise you that the files will be converted. The format used for the conversion will be the default set in iTunes Preferences on the Importing sub-tab of the Advanced tab (as seen in Figure 9.5).

After import the converted files will be ready to play.

Converting Other File Formats

Other file formats can be converted after they are in the library.

To convert a file to a specific format:

1. First choose the conversion format on the Importing sub-tab under Advanced preferences (see Figure 9.5).

2. Right-click the track you want to convert and choose Convert Selection to <format>.

Figure 9.5

Set the conversion format in iTunes' Preferences.

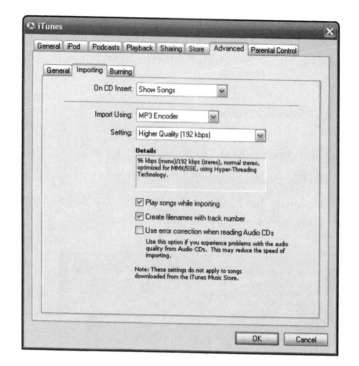

3. iTunes will convert the track immediately.

After import the converted files will be ready to play.

Winamp

In previous coverage of Winamp, I alluded to some of its abilities for playback and conversion of multiple media formats. If you are looking to convert iTunes's AAC format to WMA, here's your tool! While it cannot defeat DRM, it can be used to convert any major non-DRM file format to another.

Conversion of files is accomplished through the use of plug-ins. Encoder plug-ins impart the ability to read and write in a new format. The Transcoder plug-in gives Winamp the ability to actually perform the conversion process.

Converting AAC Files to WMA or MP3

The Transcoder plug-in can be used to convert non-DRM AAC files from iTunes to files compatible with WMA or MP3 players.

Note

The conversion process between two "lossy formats," such as AAC and MP3 or AAC and WMA, will lose a small amount of audio quality due to the fact that AAC and MP3 or WMA use different psychoacoustic principles for their encoding. While the audiophile might hear the difference, I defy him to tell the difference when riding the bus or walking down the street.

To Convert AAC files to MP3 or WMA:

1. Right-click the track or album that you want to convert and select Send to Transcode (see Figure 9.6).

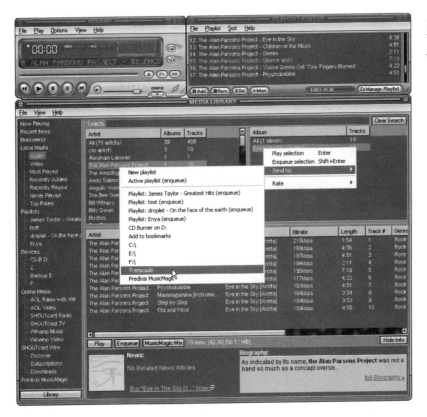

Figure 9.6

Transcoder is accessed via right-click menu option.

2. The Transcoder Configuration windows will be presented to allow you to configure Transcoder settings (see Figure 9.7). Make your configuration changes and click OK.

Figure 9.7

You can set Transcoder preferences during the conversion process.

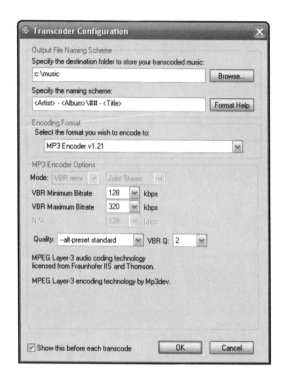

3. Transcoder will transcode the files (see Figure 9.8). If the destination folder is a watch folder (configured in Winamp preferences), the new files will automatically appear in the Winamp Media Library.

Figure 9.8

Transcoder displays a progress dialog as files are converted.

After conversion, you can play and sync the new files to any device that is compatible with your chosen format.

Note

If you choose a format such as MP3 for your converted files you will be able to play them in almost any portable media player, including the iPod.

Using Winamp with an iPod

It is possible to use your iPod without ever loading iTunes on your computer. To do this, you'll use the ml_iPod plug-in developed by Will Fisher for the Winamp media player. This plug-in connects to your iPod and syncs the Winamp Music Library content with your player's library. ml_iPod supports playlists and smart playlists, and even syncs DRM-protected content (requires iTunes to be installed, but not active).

The ml_iPod plug-in installs Transcode automatically to enable the conversion of AAC files on the iPod to MP3 or WMA and vice versa.

Windows Moviemaker

Windows Moviemaker is a complete video editing and production studio that is built into Windows XP. It is designed primarily for the production of Windows Media Video (WMV) files. It has the ability to import digital video from a number of formats, including AVI, MPEG, and MOV. The completed movies are saved in WMV format.

Converting Digital Video to WMV

Windows Moviemaker excels at the conversion of digital video formats to WMV. The process takes place in much the same way as other movie production programs. Imported video is added to a video timeline, and combined with special effects and transition effects to create a movie. The finished movie is then saved to disc or portable media for playback.

To convert an AVI video to WMV:

1. On the Movie Tasks panel, choose Import Video. Browse for the AVI file you want to convert. Click Import to begin the import process.
2. Once the video is imported into the current collection, you can place clips from it on the timeline (see Figure 9.9). If you have additional videos that you want to combine with this project, add them to the timeline as well.
3. Find and add video effects and transitions from the Edit Movie section of the Movie Tasks panel (see Figure 9.10).

Figure 9.9
Place clips on the timeline
to build the new video.

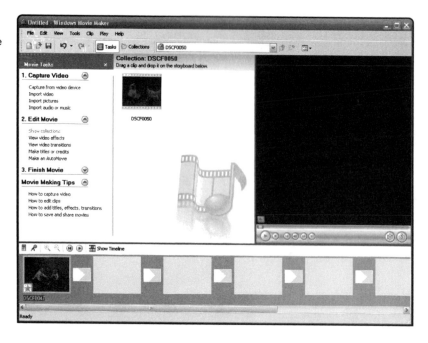

Figure 9.10
Add video effects and
transitions to add interest
to your movies.

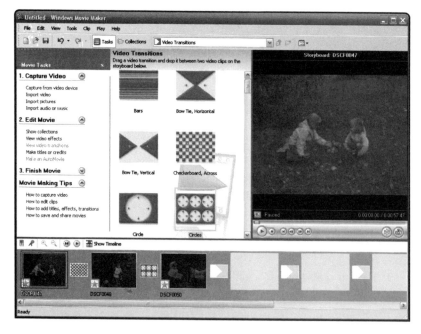

4. Choose the appropriate option under the Finish movie section of the Movie Tasks panel. The Save Movie wizard will launch to guide this process.

5. The Save Movie wizard (see Figure 9.11) helps you select the appropriate bit rate for the movie file. Higher bit rates will improve the quality of the finished movie. Complete the wizard and allow the movie to be saved.

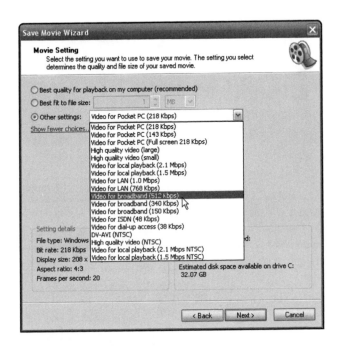

Figure 9.11
The Save Movie wizard helps you build the final movie file.

The movie will be rendered and saved in the location specified in the Save Movie wizard.

Importing Video

You can use a video capture device such as a hardware capture card or an external video capture device to import video into Windows Moviemaker. If the device is recognized by Windows as a video source it can be used to import video into Moviemaker.

To capture and produce a video:

1. On the Movie Tasks panel, choose Capture Video. Set the settings for your video source (see Figure 9.12). Click Next to continue with the Video Capture wizard.

2. The wizard will prompt you for a file name, storage location, and video settings. Complete these choices and Click Start Capture to begin the capture process (See Figure 9.13).

Figure 9.12
The Video Capture wizard guides you through the video capture process.

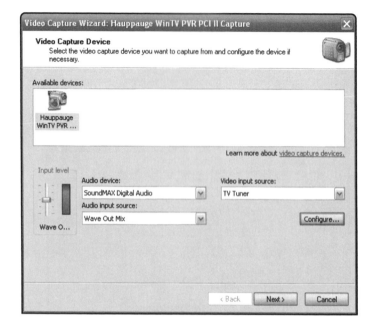

Figure 9.13
The Capture Video window lets you monitor the capture process.

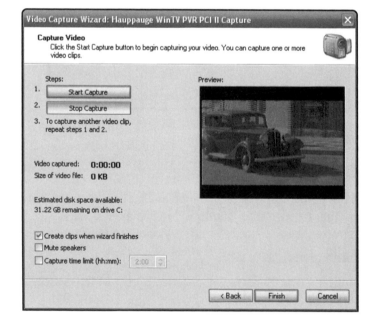

3. Click Stop Capture to stop the video capture. The captured video will be added to a video collection for use with Moviemaker.

4. Add video effects and transitions to complete the movie layout.

5. Save the movie.

Other Video Converters

In addition to the free video production capabilities of Windows Moviemaker, there are other inexpensive video conversion tools available. ImTOO offers a series of tools designed to convert between specific types of video, while SmartSoft offers a sort of one-stop-shop approach to conversion. I'll explore the main features of these tools here. Much more extensive information is available from the vendors' websites.

ImTOO MPEG Encoder

The ImTOO MPEG Encoder (see Figure 9.14) includes tools for converting files to and from WMV for playback on Portable Media Center devices and portable media players with video capability. The MPEG Encoder also includes codecs for 3GP video (for cell phones), MP4 (for Apple iPods), and even DVD formats for creation of movie discs. Operation of this converter is pretty straightforward. Choose the video to convert and the output format and go!

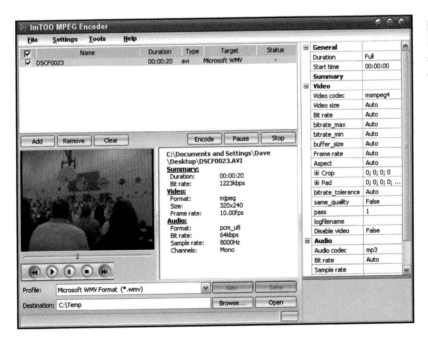

Figure 9.14
ImTOO's MPEG Encoder supports over 50 video codecs.

> **Note**
>
> You can obtain the ImTOO MPEG Encoder and other video conversion tools at www.imtoo.com.

To convert a video with ImTOO MPEG Encoder:

1. On the File menu, click Add.

2. Browse for and select the video you want to convert and click Open.

3. Using the Profile drop-down selection, choose the format you want for the converted video.

4. Click Browse to select a destination for the converted video. Choose an output destination and click OK.

5. Click Encode to begin converting the selected video.

When conversion is complete, the video is ready to sync and play in your portable device.

SmartSoft Video Converter

The SmartSoft Video Converter (see Figure 9.15) is a leaner converter. It supports fewer codecs, and outputs only to AVI, MPEG, WMV, and RealVideo. The interface is very simple and conversion speed is good. If you want a simple conversion, and do not need to support an iPod or video cell phone, this would be a good choice.

To convert a video with SmartSoft Video Converter:

1. Click Browse to select an input file.

2. Select the output format you want.

3. Click Convert to begin the conversion.

When the conversion is complete, you can sync the converted video to your portable video device.

> **Note**
>
> You can obtain the SmartSoft Video Converter at www.smartsoftvideo.com.

CONVERTING DRM-PROTECTED FILES

What I'm Listening To

Title: Red Barchetta
Artist: Rush
Source: WMA download from MSN Music
Player: Windows Media Player 10

Files encoded with a Digital Rights Management scheme will not be convertible by any normal means. Windows Media Player will transcode to lower bit rates for copying to portable devices, but will not transcode to other formats. Both Windows Media Player and iTunes will burn a CD for use in a standard CD player, but will not allow use on an unlicensed device.

It may become necessary, for various reasons, to transcode your music into another format. Perhaps you've purchased an iPod, but have music that is protected by Windows Media DRM. iTunes cannot convert this music, so you'll find yourself needing to find your own way to get it loaded into your iPod. In this section, I will present three ways to transcode music files protected by DRM. While I have not had occasion to do this myself (having both an iPod and Windows Media-compatible player), I can see the reasons one would want to do this. Heed the disclaimer at the beginning of this chapter, however. When the test of the legality of defeating DRM for personal use is made, pay attention to the results and be sure your collection conforms to any perceived changes in the law.

Converting iTunes FairPlay DRM

iTunes version 6 introduced a new version of the FairPlay DRM standard promoted by Apple. It seems that a prominent Norwegian researcher had developed a way to remove FairPlay from iTunes music and that this removal tool, called PlayFair, had been incorporated into various conversion utilities to allow users of iTunes to transcode their music for other players.

The version of FairPlay in iTunes 6 defeats this process, resulting in the effective removal of any currently known way to defeat FairPlay DRM. If you still hold songs encoded with FairPlay 5 or earlier, you can still use these tools. If you have upgraded to iTunes 6, there are other ways to convert your music, but they will involve a slight loss in quality. I will discuss the possible ways to convert DRM-protected iTunes music in this section.

Using JHymn

JHymn (see Figure 9.16) is one version of the PlayFair decoder that can be used with Windows to unlock iTunes DRM to allow the files to be converted to other formats. As stated above, iTunes 6 defeats this use of JHymn, but the tool remains effective for earlier versions of iTunes. JHymn is available from www.hymn-project.org (remember the hyphen if you don't want to get stuck in an advertising page).

Figure 9.16

JHymn removes iTunes DRM to allow use of protected music on other systems.

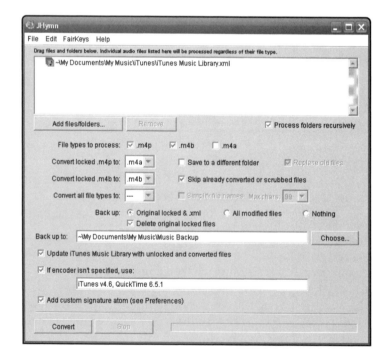

To use JHymn to remove FairPlay DRM:

1. JHymn will default to the default iTunes music folder on your system. If you want to scan other folders, click Add Files/Folders to browse for additional folders to scan.

2. Click Convert to process music files in the selected folders. As files are processed, the unlocked file will replace the original in the folder.

Converting Other DRM

The goal of a DRM removal tool is to remove the DRM component while leaving the music intact. Outside of iTunes prior to version 6, no tool currently exists to remove DRM from protected files. There are ways to rerecord your music to get an unlocked version of a file, but each involves a certain loss of fidelity. In this section I will show you how to use a CD to remove DRM from selected tracks and how to use auxiliary input recording to create unlocked copies of your music.

The CD Burner Route

Each major media player used today for DRM-protected music has the ability to create CDs from protected tracks. After burning the protected tracks to disc, you simply rip them using your favorite CD ripper. You can record them into any format you want, retaining decent fidelity, especially for portable use. This process can be used to convert protected WMA to iTunes AAC or vice versa.

To burn iTunes tracks to disc:

1. First, create a playlist containing the tracks you want to burn.

2. With the playlist displayed, click the Burn Disc button.

iTunes will burn the CD and add it to the Sources list as a music source. If you want to rip it using iTunes, just click the Import CD button to start the ripping process. If you want to rip to MP3, be sure to make MP3 your default import format in iTunes preferences.

Note

You can find extensive coverage of burning and ripping CDs in Chapter 8.

To burn Windows Media tracks to disc:

1. Locate the album you want to burn in the Library. Right-click it and click Add to Burn List.

2. The Burn List will be displayed. Click the Start Burn button to begin burning the CD.

Windows Media Player will burn the disc. The disc will be ejected when the burn process is complete. If you want to use Windows Media to rip the disc, just reinsert it and select Rip Music from CD to begin ripping the tracks to Windows Media Player's default format (configured in Windows Media Player options).

Auxiliary Input Recording

The CD burner route will offer the best possible quality recording that can be achieved outside of DRM. If you are in a hurry, or if you don't have a lot of CDs to burn up, you can record the contents of your music directly as it is played in its default media player. This process involves starting playback of the protected songs, and then recording them using a recording tool such as Freecorder or dBpowerAMP.

Freecorder

Freecorder (see Figure 9.17) is a free sound recorder that can be configured to record from your sound card's auxiliary input. This lets you record the sounds the sound card is actually making as they are made. You can record stereo in this way and the quality is not terrible (probably good enough for your morning run or a trip on the bus).

Figure 9.17

Freecorder records system sounds and music directly from the system sound card.

To record protected music using Freecorder:

1. Launch Freecorder (obtain from freecorder.com).

2. Begin playback of the tracks you want to record.

3. Click the Record button on Freecorder's control panel.

4. When you want to stop recording, click the Stop button and name the recorded track.

Note

You can record an entire album in one track if you have no need to maintain separate tracks. This will reduce the number of recording stops and starts you have to process.

dBpowerAMP

dBpowerAMp offers a similar recording feature, and also provides very good music conversion capabilities. For about $30, it will convert tracks between dozens of codecs (downloadable from dBpowerAMP's website—http://www.dbpoweramp.com/). The ability to record from the sound card is provided to round out its conversion capabilities. The maker bills its product as the Swiss Army knife of audio.

To record protected music with dBpowerAMP:

1. Launch the dMC Auxiliary Input tool (see Figure 9.18) from the dBpowerAMP start menu group.

2. If you want to be fancy, enter Artist, Album, and other information to be encoded in the MP3 tags for the recorded track.

3. Click Record to launch the recording tool (see Figure 9.19).

4. Begin playback of the tracks you are recording. Click Record on the Auxiliary Input Recorder.

5. When you have completed recording, click the Record button again to stop recording and save the track. The track is saved in the Converted Music folder on your system's main disk.

Figure 9.18
The dMC Auxiliary Input
tool.

Figure 9.19
Click Record to launch the
recording tool.

LOST IN TRANSLATION

Any transcoding process between lossy media formats will result in some loss of fidelity. Some losses will be more noticeable than others; and I defy you to tell the difference when you're riding a city bus, but it is there all the same. Audio purists will deal almost exclusively with WAV and FLAC formats, as well as the lossless versions of Apple and Windows Media codecs. Only with these codecs will you be able to ensure there is no loss of fidelity. Likewise, transcoding these formats will also retain full quality because all the information related to each sound will be there to encode into the new format.

Defeating DRM to transcode music is controversial, and not to be undertaken lightly. The DMCA forbids defeating DRM and there will eventually be high profile cases involving regular people who have removed their DRM. Until then the burn/rip method (the CD burner route) discussed in this chapter is the only reliable way to transcode your DRM-protected music. Once again, there will be loss in quality and the true audiophiles would rather just buy the music again in the new format (assuming they would be caught buying digital music at all).

Okay, go soak your head a while, get a drink of water, and meet us in the next chapter. We'll talk about how to manage this media library you are building.

the **gadget geek's** guide to

Portable Media Devices

10

Managing Your Music and Video Library

Have you ever rummaged through a box of CDs, looking for a specific album, only to grab another because you couldn't find it and you were in a hurry? Ever looked for a file on your computer that you had created weeks before, only to have absolutely no idea where it went?

Imagine these problems multiplied by the thousands. As your media library grows, you'll have to organize, or be lost. Most media players do a passable job of keeping albums organized, but fall short when you begin buying a lot of single tracks. Having a plan for dealing with this will really help you in the long run.

In this section I'll give you an overview of media library management using the top three media players. You'll see how to set up watch folders, organize your tracks with playlists, how to reduce clutter by archiving little used songs, and how to protect your media files from system crashes.

What I'm Listening To

Title: Breathe
Artist: Pink Floyd
Source: MP3 (ripped from CD)
Player: Creative Nomad MuVo NX

CHOOSE YOUR LIBRARIAN

Media players themselves make very good library management tools. Each has its strong suits and challenges. I'll present each in turn and cover basic library management features available with each. Later on I'll drill in on the tasks you'll perform with each library.

iTunes

With iPods possessing the deepest market penetration of any media player, iTunes is probably the most used media store and player. Its clean interface and close links to the iTunes storefront make buying and downloading music simple.

If there is any criticism of iTunes it is that it has almost been oversimplified for some uses, such as library maintenance. Tracks are stored in a hierarchical tree in the iTunes music folder (see Figure 10.1) and are organized through the use of playlists. These playlists can be manually created, or created automatically through the use of Smart Playlists.

Power users of iTunes will also have to live without *watch folders*, folders that other media players use to monitor for new media. Any media imports done in iTunes must be initiated manually.

Figure 10.1
iTunes music is stored in a
folder hierarchy on disk.

Windows Media Player 10

If iPods have the deepest market share in portable devices, Windows Media Player has the deepest penetration in PC media players. It has more sophisticated tools than iTunes for managing media and can be used to organize extensive media libraries.

Windows Media Player stores tracks in the My Music folder by default (see Figure 10.2). Album art and music are stored together in the same folder and an index database is stored on the system to statistics such as last time played, number of plays, etc.

Like iTunes, Media Player can use automatically created playlists for the purpose of organizing media. In addition, several folders can be set as watch folders to import new media into Media Player's central media repository.

Winamp

Winamp has been around longer than any current popular media player and has produced many of the innovations that are currently enjoyed in other media players. It can interface with both Apple and Microsoft music libraries, and can read and play anything contained therein. You'll be hard pressed to find a better digital media organizer.

Winamp's library has an extensive list of preferences and settings (see Figure 10.3). As with any computer application, with great power comes great responsibility. The configuration and management of the library will take a little time to learn how to do properly to get the most from it. I'll cover media organization and the configuration of the Winamp music library later in this chapter.

Figure 10.2
Windows Media Player uses the My Music folder by default.

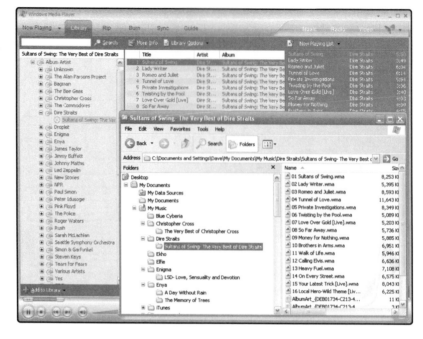

Figure 10.3
Winamp's media library is among the most flexible currently available.

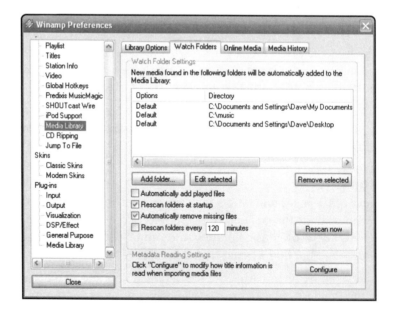

MEDIA ORGANIZATION TIPS AND TRICKS

What I'm Listening To

Title: Cliché to Your Soul
Artist: Droplet
Source: AAC (iTunes)
Player: iPod nano

As your media collection grows, you'll soon see the need to maintain organized media libraries. Organization will let you keep track of what you have and help you avoid buying any tracks twice. It will also allow you to get more enjoyment from your media; mixing it for specific moods, finding and enjoying tracks you didn't know you had, and making better buying decisions based on known preferences.

In this section I'll show you how to organize media in iTunes, Windows Media Player, and Winamp. Beyond the specifics of each of these three players, you'll also learn the basics of media management that you can use with any other player you might encounter.

Organizing iTunes

As I briefly mentioned in the introduction to this chapter, iTunes uses folders to organize media. Knowing that, you might be tempted to drop your songs into one of these folders and fire up iTunes. It's not quite that easy. iTunes keeps track of imported music in a proprietary database in the main iTunes folder. Tracks must be imported using the iTunes interface to keep this database up to date.

iTunes keeps imported media in folders according to artist and album (where possible). This allows you to quickly back up or copy your media to other systems without tedious searching. Knowing where these folders are and how to manage them can help you salvage a corrupted iTunes database and potentially save yourself a lot of money.

iTunes Storage Locations

By default, iTunes stores its music in a folder titled iTunes in your My Music folder (Windows) or your Music folder (Mac). Locating this folder is important to backing up your music, and managing any integration you might want to attempt with other media players. If you do not find the iTunes folder where you expect, you can look up the location on the Advanced tab in iTunes Preferences. Spend some time browsing this folder. Note the two iTunes database files, iTunes Music Library.xml and iTunes Library.itl. The first is a formatted text file with complete information on every track in the iTunes database. The iTunes Library.itl file stores mostly the same information (plus some user preferences information) in a more efficient format.

Importing Media

New media can be imported into iTunes using the features of the iTunes user interface. Media can be imported at any time using the iTunes Add File and Add Folder menu functions. You can copy all imported media to the iTunes folder, or you can leave it where it was found and just add it to the database. Another function, Consolidate Library, can be used to copy all externally indexed media to the iTunes folder tree to simplify backups and restoration of the library.

Adding Media to iTunes

In addition to the ability to acquire media through the iTunes music store and by importing it from CD, iTunes can import foreign media using the Add Files and Add Folders functions of the File menu.

To import foreign media into iTunes:

1. On the File menu, choose Add File to Library or Add Folder to Library.

2. A browse window will let you select the file or folder you want to add. Select the file or folder and click Open (File) or OK (folder) to begin the import.

3. If you are importing Windows Media, a confirmation dialog will offer to convert the media to the format currently set in iTunes preferences (see Figure 10.4). Click Convert to allow the import operation to continue.

Figure 10.4
iTunes will convert WMA files as they are imported, leaving the original in its original location.

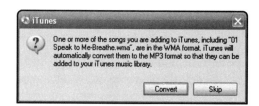

As the import progresses, iTunes will display the conversion status in the information area near the top of the iTunes window.

Consolidating the Media Library

In its default configuration, when you add foreign media that is in a format compatible with iTunes, iTunes will only create an index link to the added media. If you prefer to keep all your media organized on the file system in addition to in the database, you can use the Consolidate Library function to copy all media into the iTunes folders.

To consolidate your media into the iTunes folders:

1. On the Advanced menu, click Consolidate Library.

2. iTunes will confirm your intention to consolidate the library (see Figure 10.5). Click Consolidate to begin the consolidation process.

Figure 10.5
Consolidation copies all indexed music into the iTunes folder hierarchy.

When consolidation has completed, the consolidated tracks will be in your iTunes folders. The original version will be left where it was on your disk. If you no longer need it there, you can safely delete it.

Using Playlists

Playlists are also stored in the iTunes database. Smart playlist criteria and the contents of manual playlists are all stored there. Playlists form the basis of iTunes media organization. They allow you to group tracks by artist, genre, or by statistics such as favorites or least played. Knowing how to organize playlists will help you get control of your media assets in iTunes.

Manual Playlists

Manual playlists are good for quick track lists for playback or quickly exporting certain songs to your iPod. They are quick to set up, but labor intensive to maintain. As new media is added, the playlist remains static and must be updated manually to add or remove tracks.

To create and maintain a manual playlist:

1. On the File menu, click New Playlist.

2. A new playlist is added to the sources pane and the title is opened for modification. Type the name you want and press Enter.

3. Drag tracks from your Library to the playlist to add them. To delete tracks, display the playlist and right-click tracks you want to delete. Click Clear to remove them from the playlist.

Right-clicking on a playlist name offers you a context menu where you can choose to play the selected playlist or burn it to CD. You can also export the list to a text file in case you want a text list of the songs (I guess).

Smart Playlists

Smart playlists are what makes iTunes usable. By making a few quick selections you can organize tracks by genre, artist, rating, even size. You can set size requirements to create automatically updated selections just the right size for burning to CD. You can also use smart playlists (any type of playlist for that matter) to create iMixes for publishing your personal listening preferences on the iTunes site. Friends can use these iMixes to tweak their own preferences or as gift ideas for your next major life event.

To create and maintain a smart playlist:

1. On the File menu click New Smart Playlist.

2. On the Smart Playlist configuration window (see Figure 10.6), make selections based on your needs for this list. If you need to use multiple criteria, click the [+] sign next to the criteria line to add another.

Figure 10.6

Smart playlists can use a number of criteria to locate tracks.

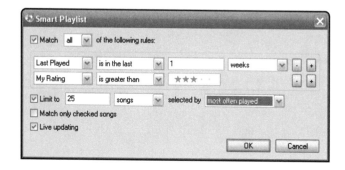

3. When you have completed making criteria selections for your smart playlist, click OK. The playlist will be added to Sources and the title opened for modification.

4. Name your playlist and it will be ready to use.

Note

If you want to create an iMix, click the small arrow to the right of a playlist title and click Create iMix. You will be directed through the process of publishing your iMix on the iTunes store site.

Built-In Playlists

Built-in playlists are a useful way to locate most played or recently added tracks. They are no more than preconfigured smart playlists. If you don't want them cluttering up your sources pane, just delete them. If you are curious about how to configure smart playlists it is often enlightening to edit a built-in playlist to see how it was created.

Using Statistics to Find Your Preferences

If you have time while you're listening to tracks in iTunes or on your iPod, you might click the little stars to indicate your rating of the track. You might listen solely on your iPod and care little for ratings. Either way, you can use statistics generated by your iPod and iTunes to create smart playlists to categorize your most (and least) favored tracks.

Do I Really Need All These Songs?

Bottom line: Will you ever listen to all 30,000 tracks in a large music library? Many of these tracks are album filler from CDs you imported, possibly long before you ever installed iTunes. You might not even know they're there. By using smart playlists you can identify tracks you've never played and deal with them. If you don't have the original CD any longer (it was damaged, right?), burn them to CD and delete them or just use the smart playlist functions to remove them from mixes you play at parties or sync to your iPod.

To remove songs from playlists or your iPod based on unpopularity:

1. Create a smart playlist to locate your little used tracks. (An example of this configuration is shown in Figure 10.7.)

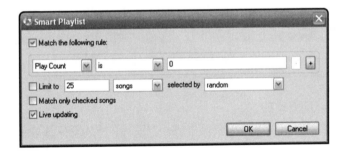

Figure 10.7

You can create a playlist to locate songs you never play.

2. Select the playlist. Choose any song in the list and press Ctrl + A (Mac use Command-A) to select all tracks on the list.

3. Right-click any track and select Uncheck Selection.

4. On smart playlists uncheck the checkbox next to Match only checked songs. This prevents the smart playlist from using the songs you just unchecked.

5. To prevent syncing the unchecked songs to your iPod, modify the iPod settings in Preferences (see Figure 10.8). Check the checkbox next to Only Update Checked Songs.

Figure 10.8
You can prevent unchecked songs from being sent to your iPod by managing your iPod preferences.

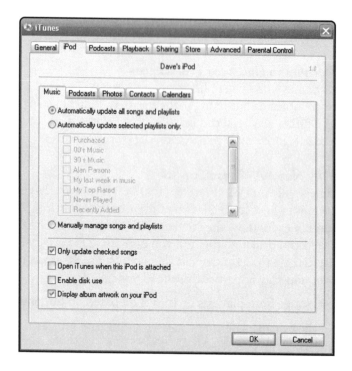

Tip

Macintosh users can achieve the right-click function by pressing the control key as you click a selection.

Removing the Songs You Never Listen To

If you just want to delete the songs from your library, you can simply burn them to CD and delete them.

To delete tracks from iTunes:

1. Select the track or tracks you want to delete.

2. Press Delete or right-click the selected tracks and select Clear.

3. iTunes will confirm your intentions (see Figure 10.9). Click Remove to remove the track.

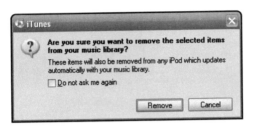

Figure 10.9

iTunes gives you one last
chance to not delete a
track from the library.

Managing Media Player 10

Media Player 10 automates much more of the process of media management than iTunes does.
With features such as watch folders and the ability to play most audio and video formats,
Windows Media Player 10 really simplifies your media experience.

There is only one major drawback to using Windows Media. If you have an iPod, you'll still need
iTunes. In addition, the songs you send to the iPod need to be in MP3 or AAC format, meaning
you cannot simply send your Windows Media files to it. You'll have to transcode them first,
potentially reducing their fidelity in the process.

Organizing Media Player Folders

Windows Media Player stores its media files in the My Music folder. Folders are organized by
artist and album, with additional information such as album art stored in the same folder.
Statistics such as number of plays and ratings are stored in a database in your user profile. Playlists
are kept in the My Playlists folder in the My Music folder. You can move Windows Media files
around on your system, but the database will have to be rebuilt to reconnect to their current
location. However, as long as you move them in a watched folder, the database will automatically
find them and change the pertinent information about the files' whereabouts.

Monitor Folders

Called monitor folders, the Windows Media Player version of a watch folder greatly simplifies
the task of managing a media library. By copying a track into a folder within the monitor folder
hierarchy, you make Windows Media Player aware of the change. It automatically locates the
files, indexes them, downloads cover art (if available), and adds the tracks to any applicable auto
playlists.

To configure monitor folders in Windows Media Player:

1. On the Tools menu, select Options to open the Windows Media Player configuration
   window.

2. On the Library tab (see Figure 10.10), click Monitor Folders to open the Monitor Folders
   configuration window.

Figure 10.10

Use Monitor Folders to watch for new media.

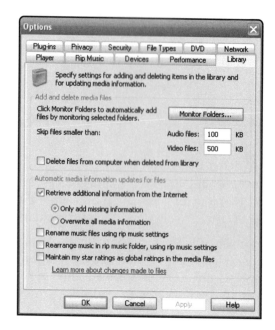

3. Click Add to add new monitor folders to the list of folders that will be watched.

4. Windows Media Player will scan the folders as time and system resources permit, adding the contents to the Library database and any applicable auto playlists.

Storage Folders

Windows Media Player uses the My Music folder in Windows XP to store media, album art, and playlists. It can also index music located anywhere on the computer. There is no Consolidate Library function like iTunes uses, so consider your storage organization plans as you begin to use your media folder.

You can move music files on your system and Windows Media Player will adapt its library to the move. When the music is moved from the old location, the music will automatically be removed from the library database. If it is moved to a monitor Folder, it will be automatically added to the database with the correct location information. This process will remove statistics such as ratings and number of plays from the database, but album art and DRM licenses will be preserved.

To move media in the Windows Media Player Database:

1. Make sure the destination folder for the media is set as a monitor folder.

2. Move the media to its new location. Windows Media Player will automatically reconcile the new location in its database.

Using Playlists to Organize Media

Windows Media Player has a very flexible system of playlists and auto playlists. Manual playlists can be quickly created and edited. Auto playlists automatically keep lists of tracks to play, burn, or sync to a portable media device.

Manual Playlists

Manual playlists are great for creating a quick collection of tracks for a special occasion. Make a playlist containing tracks for a party, date, or dinner. Add albums or songs to the playlist, save it in Windows Media Player, and burn it to a CD or sync it to your portable media device to prepare for the big event. Put your system in party mode and use the playlist to create the perfect mood for a party.

To create a manual playlist:

1. Right-click My Playlists and select New. Windows Media Player will create a new blank playlist.
2. Drag tracks and albums to the new playlist.
3. Click the small down arrow next to New Playlist and select Save Playlist As.
4. Provide a name for the playlist and click Save.

Auto Playlists

Auto playlists offer the ability to use criteria to manage media automatically. Create playlists using ratings or number of plays, genre, or artist. Tweak the criteria until you have that playlist that seems like your favorite radio station, only better, always playing the songs you want to hear without commercials!

To create an auto playlist:

1. Right-click Auto Playlists and click New.
2. The New Auto Playlist configuration window will appear. Provide a name and select the appropriate criteria for your list. (An example is shown in Figure 10.11.)
3. Click OK to save the new auto playlist.

Using Media Player Statistics

Windows Media Player maintains statistics such as number of plays or user rating on each track contained in its library. These stats can be used to help organize your tracks by making them part of the criteria for your auto playlists.

Figure 10.11

Use an auto playlist to keep content fresh and relevant to your needs.

Finding Your Favorites Quickly

If you want to see which tracks you have played the most, use the Play Count criteria in an auto playlist. Sorting by Play Count in descending order will quickly expose the tracks you have played most on your system. You can also use ratings to see which tracks you have given the highest regard. The My Rating criteria is the number of stars you have given each track.

To create a favorite tracks playlist:

1. Create a new auto playlist.

2. In the criteria selection, choose My Rating and select four or five stars.

3. Add another criteria selection to sort by Play Count in descending order.

4. Limit the number of tracks to a manageable level. Consider the number of songs you are likely to want on a CD or portable media player when you configure this setting. The results should look like the playlist in Figure 10.12.

Archiving Little Used Tracks

You can use the reverse of the preceding paragraph to locate little used or despised tracks. Delete them, or burn them to CD for archival in case your musical tastes ever change. A couple of things to look for when using this process is the fact that you have quite likely not listened to all your tracks and your statistics may not be the most up-to-date. This may happen if you have had to move media or if you have reindexed your library for any reason. Until you have a significant history of usage patterns, it may be wise to just let the songs be.

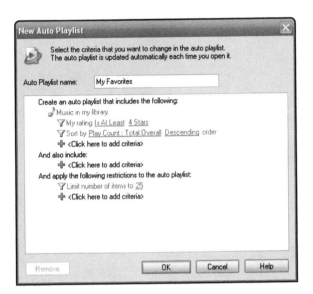

Figure 10.12
You can quickly locate your favorite tracks with the right criteria.

Wrangling Winamp

By this point, you've seen all the tricks you can use to manage your media. All that remains is to show you how to accomplish this in Winamp. Winamp does offer some compelling features that can make your media management task somewhat simpler.

Setting Up Winamp Folders

Winamp can manage media wherever it lives on your computer. If you want it to coexist peacefully with iTunes or Windows Media Player, there is no problem. Winamp can be configured to use the default media folder for your other application as its primary storage location. Use monitor folders in Windows Media Player and any song you rip in Winamp will automatically appear in Windows Media Player too!

Watch Folders

Winamp uses watch folders to monitor for new media. Rip a new song in Windows Media Player and watch it appear in Winamp. Use the best features of both media players to get more powerful control of your media assets.

To configure a watch folder in Winamp:

1. On the Options menu, select Preferences.

2. Under General Preferences, select Media Library. On the right side of the window, settings for the Media Library will be displayed.

3. Choose the Watch Folders tab to display watch folder settings (see Figure 10.13).

Figure 10.13

Watch folders let you keep Winamp in sync with other media players.

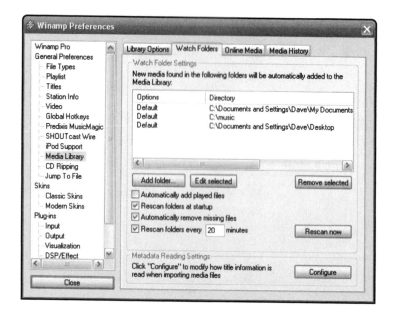

4. Use the settings on this tab to add, remove, or rescan folders. Set a rescan interval to control how often Winamp looks for recently added media.

Storage Folders

Winamp doesn't care where you want to store your music; as long as it is accessible it will be played. By use of watch folders you can ensure any media moves, additions, and changes will be automatically reflected in Winamp's media library.

Winamp Views and Playlists

Winamp uses another name for automatic playlists. Views offer the same features as those of iTunes and Windows Media Player for listing tracks based on criteria you define.

Playlists

You're probably sick of playlists by now. There is not much new to see here. I'll just quickly show you how to create one.

To create a new playlist:

1. In the Media Library window, click Library and select New Playlist.

2. Name the playlist. The playlist will be created and will be visible under Playlists.

3. To add tracks to your new playlist, just drag them to it and drop them.

Smart Views

In iTunes they are smart playlists, in Windows Media Player they are auto playlists, and in Winamp they are called smart views. These tools help you make sense of the hundreds (or thousands) of songs in your collection.

To add a new smart view:

1. In the Media Library window, click Library and select New Smart View.

2. Set criteria for your new smart view (see Figure 10.14).

3. Click OK to complete the new view settings.

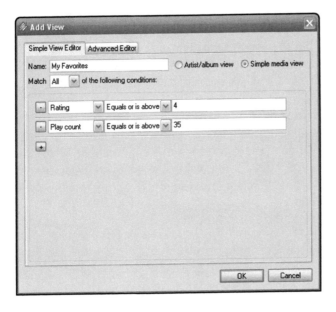

Figure 10.14
Smart views use criteria similar to auto playlists.

FILLING IN THE BLANKS

What I'm Listening To

Title: Us and Them
Artist: Pink Floyd
Source: WMA (ripped from CD)
Player: Windows Media Player 10

Your media library would be kind of flat if you only knew the names of the artists and the songs. By including descriptive information such as genre, composer, maybe even accompanying artist, you make it possible to know more about your tracks and find interesting mixtures of songs. How about only tracks where Stevie Wonder is a contributing artist? There are, surprisingly, quite a few.

Each media format has different methods of *tagging* tracks. This method of adding descriptive information, even user ratings, helps your media player find the songs you want to hear.

Tagging Your Tracks

Tagging records information about each media file in the media file itself. Each major music format has a provision for tagging. MP3s use a method called ID3 to add metadata (data about data) to the MP3 file. WMA and AAC both have their own metadata formats as well. This information can be edited manually, but the major media players can automatically retrieve metadata from online databases such as Gracenote CDDB or AMG AllMediaGuide. This metadata is recorded in the track's file and is displayed by the media player when you view track properties.

Manually Updating Your Media Tags

Each media player allows you to manually update track metadata. This is helpful when you are using tags that are not typically available in the automatic services. Windows Media Player includes a tag for mood, a very subjective rating. If you really want to have certain songs at hand the next time you're feeling mellow, though, it is a great tag to use. Set up an auto playlist for mellow tracks and turn down the lights.

Updating Tags in iTunes

iTunes calls track metadata *track information*. Right-click on a track in your iTunes library and select Get Info to display the information for the track (see Figure 10.15). Update tags, add lyrics, include artwork; you can make each track a wealth of information to help you enjoy it later.

Managing Metadata in Windows Media Player

Windows Media Player 10 includes the Advanced Tag Editor (see Figure 10.16). Configure information related to the track such as mood, genre, artist, composer, affiliated websites, and comments. You can even synchronize lyrics to the track timeline (see Turn Windows Media Player Into a Karaoke Player sidebar).

Turn Windows Media Player Into a Karaoke Player

By adding synchronized lyrics, you can display lyrics as a song plays.

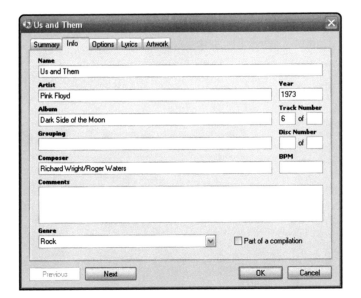

Figure 10.15
iTunes calls metadata track info.

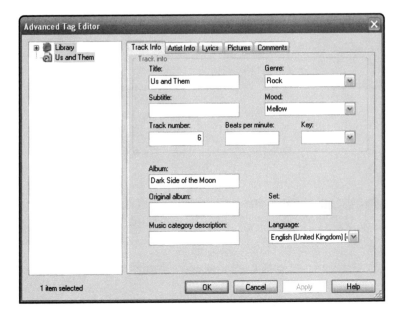

Figure 10.16
The Advanced Tag Editor lets you add metadata to WMA files.

To use the Advanced Tag Editor to add lyrics to the timeline of the song:

1. Open the Advanced Tag Editor and select the Lyrics tab.

2. Click the Synchronized Lyrics button to open the lyrics editor (see Figure 10.17). A time-line of the song will be displayed.

Figure 10.17
Use Synchronized Lyrics
to create lyrics that display
during playback.

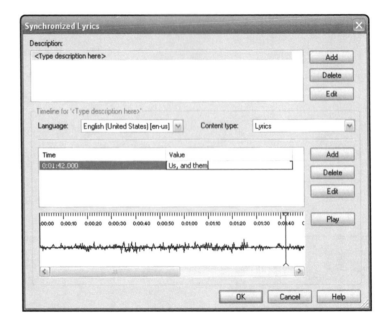

3. Click Add to add a lyric to the timeline. Type the lyric text and give it a time code.

4. Add any other lyrics and click OK to save the synchronized lyrics. Click OK to close the editor.

To Display Synchronized Lyrics

On the Play menu in Windows Media Player, choose Captions and Subtitles and select On if Available. Lyrics will display just below the visualization area in Windows Media Player (see Figure 10.18).

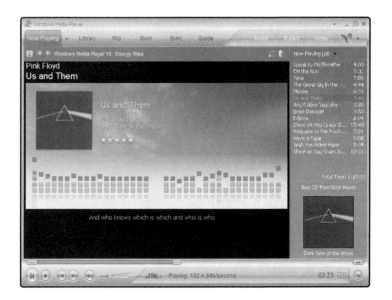

Figure 10.18
Synchronized lyrics
display during song
playback.

Tagging Tunes in Winamp

Winamp cannot modify more than basic information on WMA and AAC media files, but has more extensive options for editing MP3 tags (see Figure 10.19). It supports ID3V1 and ID3V2 (a newer version of the ID3 tag) tagging to be compatible with most current MP3 tagging schemes.

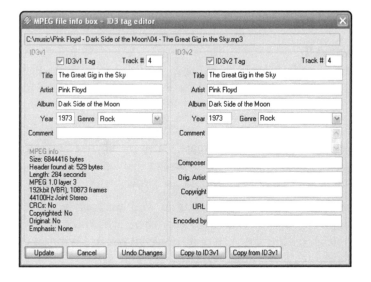

Figure 10.19
Winamp supports both
ID3V1 and ID3V2 tags for
MP3 files.

Automatically Retrieving Metadata

iTunes and Windows Media both fill in the blanks of ripped or imported tracks using online media databases. Winamp will import any metadata it finds in imported music files, and will obtain online track information from Gracenote CDDB during CD playback or ripping.

CONSOLIDATING MUSIC FROM ALL SOURCES

What I'm Listening To

Title: Silence Must Be Heard
Artist: Enigma
Source: MP3 (ripped from CD)
Player: Winamp 5.11 Pro

Imagine for a moment that you have two or more portable media devices (one of them an iPod), maybe a PlayStation Pro, a couple of PCs, and a car stereo that plays WMA and MP3. If you want to be able to listen to your music on each of these devices, what is the best way to manage your media? If you buy music in iTunes can you play it in your car? If you go all out for quality, say with Apple lossless codec for ripping music from CDs, can you play it anywhere but in iTunes and on your iPod?

In this section, I'll pose some of these questions and show you real ways to unify your media holdings and enjoy them regardless of which device plays them.

One Format to Rule Them All

Pop quiz: Which media format we have discussed in this book can play on an iPod, a PlaysFor-Sure compatible portable media player, and in all three of the major media players we have discussed?

If you answered MP3, you are absolutely right. For all the whining the format purists do about the MP3 format, it remains the lingua franca of media formats. If you rip all your CDs into MP3, you can be assured the songs will play anywhere and in all devices. Go for high fidelity with either Apple or Microsoft and you begin to paint yourself into a corner.

Now, if you have only an iPod and play tunes in the car using an iPod link, play songs at home on an iPod external speaker set, and sprout white earbuds everywhere else, you can probably get away with iTunes and the Apple media formats. You're set as well if you use Windows Media Center Edition, a PlaysForSure portable player, and a WMA-compatible car stereo.

Configuring a Single Media Source

The task of configuring one media library for all media players and media devices is not as difficult as it seems. iTunes and Windows Media Player 10 both use the My Music folder for storage of media. Winamp doesn't care; it will use any folder you configure in Preferences. If you set Winamp to use My Music as well, all three meet in the middle, and can definitely coexist on the same system. Each varies with the type of database used, but all use the same folder structure to store music.

Configure the My Music folder as a watch folder to ensure all applications are able to keep their libraries' databases up-to-date. If you rip a song in Media Player, it will be automatically added to Winamp. A new MP3 track in iTunes will appear in Windows Media Player.

Note

iTunes does not support watch folders. Any tracks added to the libraries of other players will need to be imported into iTunes manually.

Keeping all your music under one lid takes a little coordination when you install the various media players, but is fairly simple. With both iTunes and Windows Media Player finding their own uneasy truce, the folder structure is in place for Winamp to use.

To configure Winamp to use the My Music folders:

1. Configure Winamp to watch the My Music folder for new media.
2. Set the CD ripping settings to use My Music as the default output folder (see Figure 10.20).
3. Set default folder settings in any plug-ins that import or export media.

Once these settings are configured, Winamp will scan the My Music folder tree and index any available tracks in the Winamp Music Library. These tracks are then available for playback in Winamp or syncing to any portable media player compatible with Winamp.

Note

With appropriate plug-ins Winamp can sync to both iPods and other portable media players. DRM-protected media will have to be synced using the native media player.

Figure 10.20
You can configure the default folder used for ripped tracks.

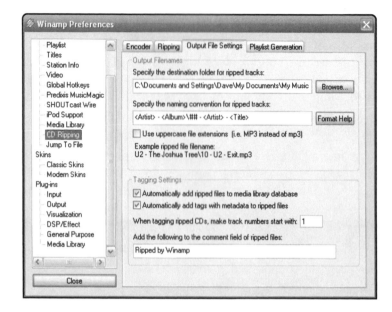

BACKING UP YOUR MEDIA CONTENT

What I'm Listening To

Title: Still Crazy After All These Years
Artist: Paul Simon
Source: WMA (ripped from CD)
Player: Windows Media Player 10

If you have ever had to reload a computer, you know that there is a very good chance that your digital media files will be lost some day. If you've acquired them online, there may not be a CD version of the file to restore to your system after you get it back online.

In this section I'll show you how to backup and restore your media files to ensure that you can quickly recover your library if disaster should strike.

Why Back Up?

If you don't relish the thought of sitting in front of your computer for hours feeding it CDs, or downloading several hundred songs from the music store again, you'll see the value in creating a simple backup of your media files. With proper backup and recovery preparation, you should be able to restore your media in a matter of minutes after your system is recovered.

Disk Backups

The fastest way to back up media files is to make a simple copy to another disk on your system. This allows you to very quickly copy them back if you run into problems. You might have to install a second disk into your system to allow you to use this option. The cost of this move shouldn't be much more than $100.

This solution will not protect your media against a catastrophic event in your computer. If the disk the media is saved to is damaged, you will lose your backup copy of these files.

External Disk Drives

External disk drives solve the catastrophic failure worry, and are only slightly more expensive than comparable internal disks. Many also come with pushbutton backup programs that help you create system recovery backups for even quicker recovery of your system and files.

CD-R Backups

If your media collection is modest you can back it up on one or two CDs. Using your system's built-in CD burning capabilities you can copy your files over to CD-R and burn them to disc. Store the CD-Rs in a safe place. If you have to restore your library, just copy the files back from the CD.

Portable Media Devices

Many USB keychain drives can be used to backup a limited number of media files. iPods and many portable media players can also be accessed as file storage and can be used in a pinch to backup or transport media files.

Be sure your portable device appears to the system as a disk drive. Copy files to the device, being careful not to overload the device. Recovery is as simple as copying the files back from the device after the system is restored.

DIGITAL RIGHTS MANAGEMENT AND TRANSPORTING MEDIA

What I'm Listening To

Title: Yours Is No Disgrace
Artist: Yes
Source: WMA-DRM (From MSN Music)
Player: Windows Media Player 10

The backup and recovery steps above will be a bit more challenging if you are dealing with protected media. DRM prevents you from playing the media on a system that doesn't have a license to play it. It is important to back up this license to ensure any recovery will include the DRM-protected tracks as well.

Moving Your Media to Another Computer

When you move your media to another computer, you'll want to enable the destination computer to play the protected content. This can be done in one of two ways. You can back up the license keys and restore them to the destination computer or you can recover the key from the music store where you bought the music. Windows Media Player can use either method to recover keys. iTunes uses a slightly different process to manage protected music. Each iTunes system is *authorized* by the iTunes music store. A limited number of systems can be authorized. To recover a license, you simply need to authorize the system you are using.

Backing Up and Restoring Your License Keys

Windows Media Player can back up license keys to disk to enable the ability to recover DRM-protected tracks. If you have these keys, you do not need to connect to the store where you purchased the tracks.

To back up your DRM keys:

1. On the Tools menu, click Manage Licenses.

2. In the Manage Licenses window (see Figure 10.21), click Change and select a location to store backup licenses.

3. Click Back Up Now to back up your licenses.

Figure 10.21

Use Manage Licenses to back up and restore licenses in Windows Media Player 10.

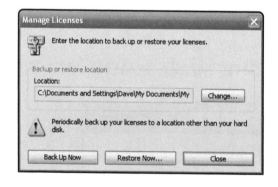

To restore a DRM license key:

1. On the Tools menu, click Manage Licenses.

2. Use the Restore Now button in Manage Licenses to restore your licenses.

Tip

You can restore Windows Media licenses backed up on another system to move them to a new system or recover a downed system.

Recovering a Lost DRM License Key

If you have not backed up your DRM keys, all is not lost. You can usually recover lost keys from your online music store. They keep track of the media you have purchased and can reauthorize you to play it. Check with your media store to see what their process is for re-creating your licenses.

NOW THAT YOU'RE ORGANIZED…

Hopefully this chapter has given you some ideas about how you'd like to arrange your media library. You may not be so hardcore that you'll be using three or more media players and several portable devices, but it is important to know that you have options and that you can migrate from one device to another without leaving some or all of your media behind. The perception of incompatibility is often the tactic used by a dominant market player to keep its customer base in line. It has been used to good effect in many industries and shrinks away in the light of a thorough examination of the other options.

Try out some of the things you've learned here. Install and use Winamp to play iTunes media. Backup your media so you can finally do that hard reset you've been afraid to do on your iPod. Organize your tags so it's easier to create playlists that reflect your many moods.

That's about it for the portable media learning. From here on out we'll have fun with podcasts, both listening and creating, and I'll give you a glimpse of the future of portable media devices.

the gadget geek's guide to
Portable Media Devices

11

Podcasts; Not Just for Ubergeeks Any More

If you have read much about iPods and other portable media devices, you've read about podcasts and podcasting. The best thing since early man learned how to slice bread (apparently), podcasting has actually been around long before it had a name. Many claim to be the father of modern podcasting (ever notice it's never "the mother"?). Among those mentioned are Adam Curry (former MTV Vee-Jay and podcast aficionado), Userland.com founder Dave Winer, and Userland.com user Tristan Louis. The reality lies someplace in between, with many folks sharing in the creation; sung and unsung heroes to those who now enjoy daily podcasts on a variety of topics.

In this chapter I will cover all things podcast. From history to future—Apple to ZiePod—you'll learn how to locate, acquire, and download podcasts; how to get them on your portable media device; and what developments to expect as podcasting continues to evolve.

PODCAST BASICS

What I'm Listening To

Title: Your Mama Don't Dance
Artist: Loggins and Messina
Source: MSN Radio
Player: Windows Media Player 10

The term podcast is a portmanteau, combining parts of the words iPod and broadcast. The term, as near as anyone can tell, was first used publicly by Ben Hammersley in an article for *The Guardian*, a British Newspaper, in February of 2004. While podcasting relies neither on an iPod nor on any actual broadcasting mechanism, the term stuck. Proponents, in an effort to clarify podcasting's true nature, have also referred to it as personal audiocasting (paudcasting?) and personal, on-demand casting. Whatever the etiology, podcasting has become a popular and personalized way to obtain news, information, and entertainment.

What Is a Podcast?

A podcast is basically a digital media file that is advertised through one or more means to client software that is designed to find and download it. The first podcasts used Really Simple Syndication (RSS) feeds published by weblogs (blogs) that advertised the existence of the media files. The Radio Userland blog reader was designed to locate and download these feeds to the user's media player for offline listening.

The History of Podcasting

In 2001 Dave Winer, the founder of Userland, responded to requests from users Tristan Louis and Adam Curry to add the ability to advertise audio feeds as part of their blogs. Readers of the blog would have the opportunity to download and listen to these feeds within the Userland application. The technology was extended to begin automatically downloading these audiocasts to the client for offline listening and later to synchronize these files to the listener's portable media player.

Pre-Historic Podcasts

Way before podcasting was cool, all the pieces were there to build it. Many people placed MP3 files on their personal websites, Internet newsreaders had been automatically downloading discussion posts for over ten years, and several technologies existed to make folks aware of new content on websites and blogs. New audio files were advertised through these mechanisms, and often were included as attachments to these posts and notifications. All that was required was the bright spark that brought it all together.

First Podcasts

Userland users Tristan Louis and Adam Curry (of MTV fame) both separately requested founder Dave Winer to extend his blogging service to include the ability to automatically distribute audio files along with blog posts. Without the catchy name, these posts became the first podcasts. Listeners at that time could hear these audiocasts by playing them on their computers. Those with MP3 players could copy audiocasts to it for offline listening, but no mechanism existed to perform this automatically.

At an Internet blogging conference in 2003, Kevin Marks demonstrated an Applescript application for Macintosh computers that automatically synced blog audiocasts to iTunes (and by extension, to an iPod). Adam Curry expanded this idea, creating Syncpod, which became the basis for iPodder, the first widely available *podcatching* software. Podcast listeners now had the complete toolset needed to obtain and listen to podcasts.

The Podcast Explosion

Podcasting soon became available on other computers as developers jumped on the concept, creating podcatching software for Windows, Linux, and various versions of UNIX.

More and more bloggers began using podcasting as a tool to liven up their blogs and expand the types of content they were able to offer. After realizing it was going to happen with or without them, Apple included podcasting as part of iTunes 4.9 for both Macintosh and Windows.

Podcasting continues to grow exponentially. Beginning with a total of 24 hits in September 2004, a Google search for "podcasts" in December 2005 finds over 113,000,000 hits!

Podcast Clients

Several podcatchers are available as stand-alone applications for finding and downloading podcasts to your computer. Among these is Juice, the free podcatcher originally conceived (as iPodder) by Adam Curry and donated to the open source development community. iTunes now also integrates podcast retrieval, and methods exist to integrate podcatching into Windows Media Player and Winamp, rounding out the big three media player applications we have been exploring.

iTunes

iTunes includes built-in podcast support. Using the podcast directory located on iTunes.com (see Figure 11.1), you can browse and select from thousands of available podcasts. Other podcasts can be subscribed to directly by simply entering the URL of the podcast syndication file.

Figure 11.1
iTunes integrates a podcast directory directly into the application.

iTunes will synchronize podcasts directly to your iPod, allowing you to take them on the road. Bear in mind, though, that many podcasts are over 30MB and last over an hour. Be sure you have room on your iPod to accept the contents of the podcast.

Windows Media Player 10 (with Help)

Windows Media Player 10 does not have direct podcast subscription ability. If you want to synchronize podcasts using Media Player, you'll need to use the Monitor Folder function of Media Player to discover podcast media files downloaded by another podcatcher. Once discovered, these podcasts can then be played in Windows Media Player (see Figure 11.2) or downloaded to your portable media device.

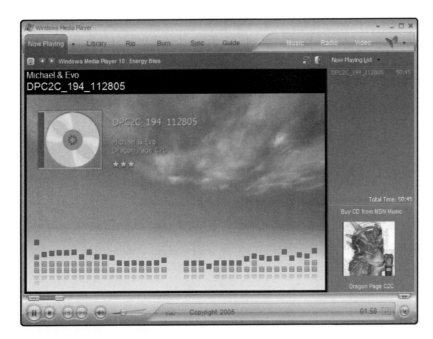

Figure 11.2
Windows can play, but cannot catch podcasts.

Later in this section I will show you the applications that help Windows Media Player automate the podcatching process. It is possible, with the help of these applications, to not only acquire, but to automate the synchronization of these media programs to your portable media player.

Winamp (with Help)
Winamp is able to acquire podcasts without help from a dedicated podcatcher. Once "caught" however, the podcast must be manually synchronized to your portable media device. Winamp uses the SHOUTcast Wire service, a directory of podcasts published by the AOL SHOUTcast service. Using Winamp menus (see Figure 11.3), you can discover, subscribe to, and download podcasts. Once downloaded, your podcasts can be dragged to your portable media device for offline listening.

The "Help"
Several applications exist for the purpose of catching podcasts. I will evaluate three free podcatchers in this section, showing you how they differ and how they can be used to aggregate podcast content for use in Windows Media Player and any other media player capable of using watch folders.

Figure 11.3

Windows can subscribe to and play podcasts using the SHOUTcast service.

Juice

Formerly iPodder (until the Cease and Desist order arrived from Apple), Juice (see Figure 11.4) is the original open source podcatcher conceived by Adam Curry and Dave Winer. Juice manages the process of locating and subscribing to podcasts.

Figure 11.4

Juice was the first real podcatcher.

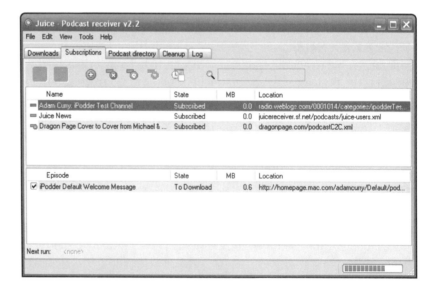

Juice is free to download and use, and is supported by a group of volunteer programmers. It is *open source*, meaning the program code is open to anyone who wants to see it and use it to create new features or similar products. As is characteristic of many open source programs that receive frequent updates, Juice has the ability to check for updated versions. These versions may fix bugs in the program or add new functions to the program.

Juice can be configured to download podcasts to a certain folder, a good feature to use if you plan to use Windows Media Player's Monitor Folder technology or Winamp watch folders. It can also play podcasts by passing the filename to Windows Media Player or Winamp for playback.

Juice has Windows and Macintosh versions and is working on a Linux version.

Doppler

Doppler (see Figure 11.5) is also a free product supported by volunteer programmers. With a feature set similar to that of Juice, Doppler allows one to locate and download podcasts to your media watch folders. Doppler can also read blog entries that may be included in the podcast feed to allow you to use the same program to monitor blogs and listen to podcasts.

Doppler is designed for Windows only.

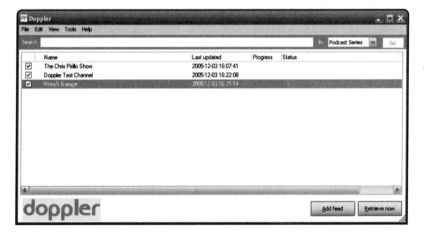

Figure 11.5
Doppler can synchronize podcasts to media watch folders.

ZiePod

ZiePod (pronounced with a long "e") rounds out our tour of three free podcatchers. Most of the features of the other podcatchers are present in this version as well (see Figure 11.6). There are a few features that help it stand out from the others. One is the ability to manage exports of podcasts to portable media within the ZiePod application.

Figure 11.6
ZiePod adds some inter-
esting usability features to
the standard collection of
features.

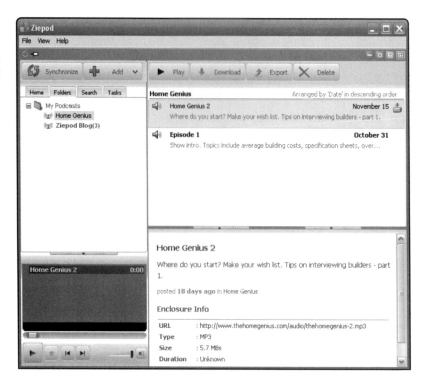

Another unique feature of this podcatcher is the ability to display four different modes, including a very compact version of the user interface (see Figure 11.7), during playback using a small amount of the screen while still displaying critical statistics about the podcast. To toggle between these modes, just click one of the colored square icons on the ZiePod toolbar.

ZiePod is available for Windows only.

Figure 11.7
ZiePod has a very handy
toolbar mode that gets it
out of the way.

Mobile Podcast Players

I've covered the major podcast acquisition tools. Let's look now at the ways you can take your podcasts on the bus. So far in this book we have concentrated on portable music players when we've talked about portable media players. For podcasts let's also look at devices that might be used to acquire podcasts without a computer.

iPods and Portable Media Players

Okay, well, the first stop should be the digital media players we have come to know and love. If you are listening to a music podcast, these will be your first choice of course. They offer the best fidelity and performance and are designed expressly to play massive sound files such as those you will be downloading from podcasters.

Mobile Phones

Mobile phones can receive podcasts as MMS (Multimedia message) or J2ME (Java MIDIlet) messages. This requires special software on the part of the podcaster, but is an effective way to communicate with mobile audiences.

On the client side, Mobilecast (see Figure 11.8) is an extension to Juice that allows you to transcode podcasts to media files using the adaptive multi-rate (AMR) codec. This allows AMR-compatible phones to play podcasts that you have loaded in from your computer.

Figure 11.8
Mobilecast prepares podcast audio files for use on AMR-compatible mobile phones.

Mobilecast is launched automatically after each podcast is downloaded to split the podcast's MP3 file into AMR files, each having 10 minutes of the main podcast content.

Personal Digital Assistants

Many Personal Digital Assistants (PDAs) have built-in media players that can play MP3 files. The Windows Mobile-based PDAs include Windows Media Player Mobile. This version of Windows Media Player can automatically sync with Windows Media Player 10 to receive podcasts. Play-back of the podcasts is accomplished using Windows Media Player Mobile on the PDA.

Palm-based PDAs can also play MP3 files. These can be synchronized by copying them to the Palm using the Palm Desktop or third-party sync software such as Intellisync. The actual podcast can be acquired using ZiePod, Juice, or Doppler and synced using Intellisync, which can be configured to automatically synchronize certain folders with the Palm.

USING PODCASTS

What I'm Listening To

Title: Who Can it Be Now?
Artist: Men at Work
Source: Club 80s w/DJ Lex on Club 977
     (SHOUTcast Radio)
Player: Winamp Pro 5.11

Well, we've had the quick tour. Now it's time to learn how to do all this. In this section, you'll learn the basics of locating podcasts, subscribing to them, and downloading them. You'll also learn how to sync them to your portable media device to take on the road.

Finding Podcasts

Most podcast producers do all they can to be noticed. This is great because it makes your job easier. Most are listed in one or more of the major podcast directories. If you don't find what you're looking for here, a quick web search will uncover thousands (including many of those you saw in the directories).

Begin with the directories loaded into your podcatcher. These are verified by the podcatcher to be compatible with its programming. If you don't find what you need here, you can access the directories, well, directly. Browsing the directories will lead you to the podcast's home page where you will have an opportunity to subscribe to the podcast.

Podcast Directories

The major podcast directories provide a place for podcasters to advertise their services, connecting with the greatest possible number of interested individuals. They are a two-edged sword at times, inundating prospective listeners with choices, but are invaluable to the consumer who might not otherwise find a particular podcast.

The original podcast directory, iPodder.com lost its trademark fight with Apple and changed its name to indiepodder.org. It remains a principal source of podcast listings, but has been passed by sites with more sophisticated indexes.

Podcast Alley

Podcast Alley is a great place to begin podcatching. With Top 10 and Top 50 lists, you can quickly subscribe to the top podcasts as voted on by Podcast Alley listeners. Subscribe to the top list individually or use the Top 10 and Top 50 feeds to grab the entire Toplist!

To subscribe to Podcast Alley feeds, browse or search for feeds using the podcast search tool. Click the Subscribe link next to the feed description. When the feed URL is displayed, copy and paste it into your podcatcher.

PodcastPickle.com

Billed as the first podcast community, PodcastPickle.com (see Figure 11.9) has one of the most comprehensive and entertaining podcast directories in existence. With over 4,000 individual podcasts listed, there is something here for every taste.

Figure 11.9
PodcastPickle.com provides an entertaining directory of podcasts.

To find a podcast on PodcastPickle.com, use any of the following methods:

- Enter a search keyword in the Search the Pickle box and press Enter.
- Browse the categorized podcast lists.
- Check out the featured podcasts listed on the site's home page.

Feedzie

Feedzie brought folksonomy to the world of podcasting. Folksonomy is the system of tagging used by sites such as del.icio.us to create user-maintained directories of information, organized by the relative popularity of each tag. By assigning tags to podcasts, podcast listeners join in the categorization and ranking made possible by this technology. Each tag can be displayed in a map with a size and weight proportionate to its popularity (see Figure 11.10). This lets you intuitively see which genres are more popular and directs your search to the more popular (more highly tagged) categories. This method of classification relies on the ability of a population to detect listening trends better than highly paid advertising executives.

Note

Feedzie is reluctant to use the term *podcast* so you will see references on this site to "audio feed" instead.

Figure 11.10
Tagging makes folkso-nomic directories possible.

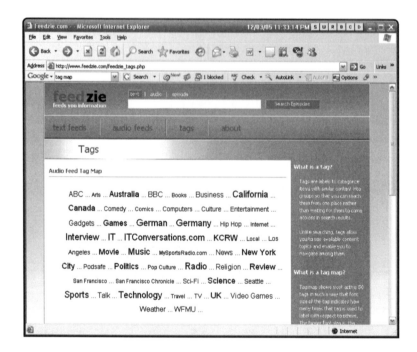

To use Feedzie to locate podcasts:

- Click Audio Feeds to open the podcasts directory and browse the categorized listings.
- Click Tags to browse podcasts by popularity.
- Use the Search box to search audio feeds (podcasts). Be sure to click Audio Feed just above the search box to search the audio feed listings.

Yahoo! Podcasts

Yahoo! brings its search engine power to bear on the podcasting world! This directory also makes use of the Feedzie-style tag maps to aggregate podcasts by tag and to present a graphical representation of the relative popularity of each tag category or genre.

The Yahoo! Podcasts directory also works with the Yahoo! Music Engine (see Figure 11.11), a media player designed to interface iTunes-style with Yahoo! Music. By using the Yahoo! Music Engine users can download podcasts listed in the Yahoo! Podcast directory to their computers and portable devices.

Figure 11.11
The Yahoo! Music Engine is a full-featured media management system.

To locate podcasts by using the Yahoo! podcast directory:

Install the Yahoo! Music Engine and the Podcast Directory extension (provided free by Yahoo!).

- Search for podcasts using the keyword search function.
- Browse listings by category.
- Browse tagged listings.
- Browse featured listings.

> **Tip**
>
> Use the tagged listings to find podcasts organized by what the podcaster thinks the podcast is about. Use the category lists or feature lists to locate podcasts according to what Yahoo! thinks they are about.

iTunes

Ahh, iTunes! With its integrated podcasting features, and the obvious connection between the term podcast and the iPod, you'd think they had invented podcasting. In fact, iTunes did not have native podcatching ability until version 4.9 was released in the summer of 2005. That said, when Apple does something, they do it right. By November 2005, over 7,500 podcasts had been indexed in iTunes and made available to users of iTunes and the Apple iPod. Downloads are automatic, and are automatically updated to the iPod when it is connected to the computer. Leave your computer on and your iPod connected and you'll have new podcasts for your morning commute!

To locate podcasts with iTunes:

1. In the Source pane, select Podcasts.
2. Near the bottom of the iTunes window, click the Podcast Directory link.
3. Search or browse the podcast listings in the iTunes Music Store.
4. When you locate a podcast you want to subscribe to, click the Subscribe button.
5. Confirm your intention to subscribe if prompted by iTunes.

iTunes will update subscribed podcasts and sync them to your iPod for offline listening.

Search Engines

Search engines are a great way to locate podcasts. With the prodigious use of certain search engine tricks, you can narrow down the list to a few good selections very quickly. If you haven't looked into search engine syntax, spend a little time researching the keywords and symbols used with Yahoo! and Google to tweak searches. You'll be glad you did.

Here are a few tricks you can use right away to locate podcasts using Google:

- Use the term "podcast" to immediately restrict hits to pages using the word podcast.
- Use quotes around terms you want to see used together. (Example: "big wave surfing")
- Use the plus sign in front of any common words such as *the* or *and* to tell Google that any returns must include that keyword.

Use a minus sign before any word you want to exclude.
(Example: use –Hawaii to exclude podcasts about Hawaiian big wave surfing.)

Blogs

If you already subscribe to a blog that has a podcasting component, just subscribe using your favorite podcatcher. You can locate additional blogs on topics you find interesting by using services such as Technorati.com or Blogger.com.

Downloading Podcasts

Locating the podcast is only the first step to getting it onto your system. In the previous section you may have noticed that iTunes was the only service where I listed the steps to subscribe. This is because it is literally a one-click operation. In this section I will describe the methods used to subscribe to each podcatcher and will also cover major configuration settings on each application.

Juice

As the descendent of the very first podcatcher, Juice should have the simplest user interface and best features for locating and subscribing to podcasts. Well…

As long as you want to subscribe to podcasts listed within Juice's built-in directory you are all set. These directories come from the indiepodder.com podcast listings, and are pretty complete, but are by no means all-encompassing.

To subscribe to a podcast feed listed within Juice:

1. On the Podcast Directory tab, click on a Directory heading to download the listings under that heading.
2. Click the plus sign next to the heading you are browsing.
3. Click on a podcast you want to add (see Figure 11.12) and click the Add button.

To add a feed not listed in Juice:

1. Locate the URL to the feed on the Web page for the podcast.
2. Press Ctrl + N or click Tools and select Add a Feed.
3. Enter the feed URL in the URL field of the Add a Feed window (see Figure 11.13).
4. Click Save to add the feed.

Figure 11.12
Browse the Podcast Directory for podcasts listed on the indiepodder.com directory.

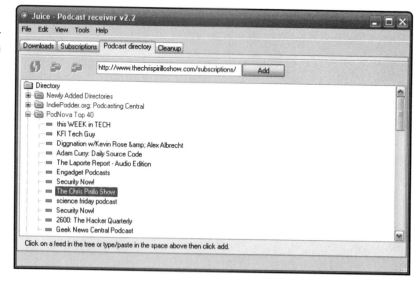

Figure 11.13
Use the Add a Feed feature to add non-listed feeds.

Once a feed is added, the episode list will be downloaded from the feed's syndication file. You can check the checkbox next to each episode and click the Download button (see Figure 11.14) to download the podcast to your system. Once downloaded, you can listen to them or sync them to your portable media device using Windows Media Player or Winamp.

Doppler

Doppler works in much the same way as Juice. With a limited built-in directory to draw on, you'll need to add more podcasts manually. Once added, the feeds can be downloaded and played. Doppler doesn't have a built-in ability to play podcasts, but uses Windows Media Player or Winamp to play and sync the podcasts.

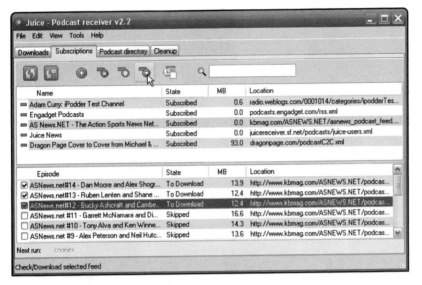

Figure 11.14
Select and download any episodes that you are interested in.

To add a podcast feed to Doppler:

1. Click the Add Feed button to open the Feed window (see Figure 11.15).

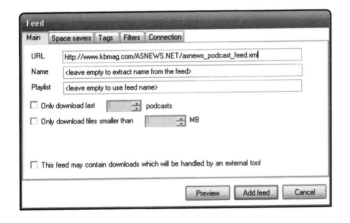

Figure 11.15
The Feed window allows you to configure podcast feeds.

2. Enter the URL of the podcast feed into the URL field.

3. Click Add feed to subscribe to the feed.

4. In the main window (see Figure 11.16), click Retrieve now to download the feed to your podcast folder.

213

Figure 11.16
Doppler can retrieve multiple feeds at once.

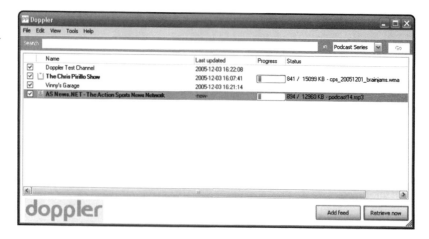

Tip

Doppler hides some amazing features in the right-click function of your mouse. By right-clicking a feed, you can open a list of all posts in the feed, read text information related to each post, even download the audio attachment if it has not already been downloaded. You can *catch-up* the feed (or signal Doppler that you want to flag it as current). You can even preview podcasts with a right-click.

ZiePod

ZiePod has two distinct subscription mechanisms. When using the Feedzie podcast directory, you can use a built-in function designed specifically to use Feedzie links. (That's where the "zie" in ZiePod comes from). You can also subscribe to feeds in the classical way.

To subscribe to a Feedzie.com podcast feed:

1. On the Feedzie.com site, locate the feed you want to subscribe to (see Figure 11.17).

2. Click on the + sign next to the feed.

3. If ZiePod prompts you to confirm the selection, click Add to complete the subscription.

To add a new podcast feed manually:

1. On the ZiePod main window, click Add and select Add Podcast.

2. In the Adding a New Podcast window (see Figure 11.18), enter the URL for the podcast feed.

3. Click Finish to complete adding the new feed.

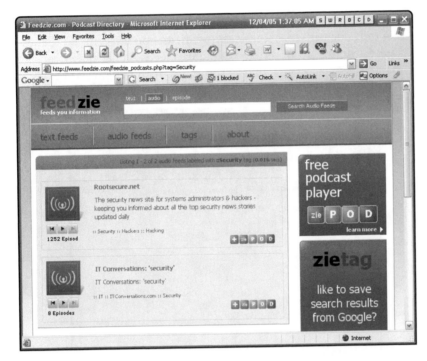

Figure 11.17
ZiePod offers one-click subscription from the Feedzie podcast directory.

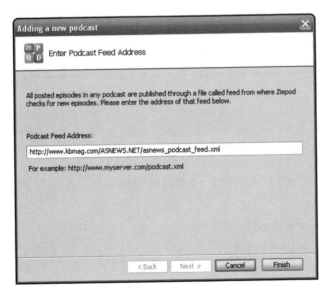

Figure 11.18
Adding a new feed manually isn't that hard, either.

Once all the feeds you want are added, you can use the Synchronize button on the main window to download podcasts to your system.

iTunes

You almost fall over podcasting in iTunes; it's that simple. Browsing the podcasts in the iTunes Music Store you simply click on Subscribe to subscribe to a podcast feed. iTunes downloads podcasts according to your preferences (see Figure 11.19) and syncs them to your iPod.

Figure 11.19
You can manage your podcast schedule and retention policy in iTunes.

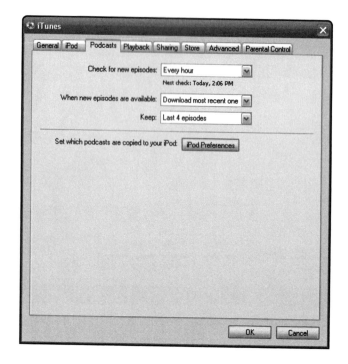

Winamp

Winamp uses the SHOUTcast Wire service to locate and download podcasts. Podcasts are delivered to a designated download folder, and added to the list of downloaded podcasts in the SHOUTcast Wire section of the Music Library index. Podcast subscription settings are configured in Winamp Preferences in the SHOUTcast Wire settings section (see Figure 11.20).

Syncing a Podcast to Your Portable Device

Once you've subscribed to a variety of podcasts, you'll want to take them out for a spin. You can sync them to any device compatible with the media format used in the podcast. Since most are in MP3 format, you'll be able to use them with virtually any device you choose. In this section, I'll show you how to use iTunes, Windows Media Player, and Winamp to sync your podcasts. I'll also show you how to sync podcasts to a Pocket Windows device by using Microsoft ActiveSync.

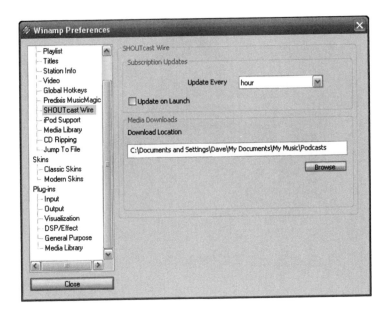

Figure 11.20
You can configure pod-
cast download schedules
in SHOUTcast Wire
settings in Winamp
preferences.

iTunes

You can manage the synchronization process between iTunes and your iPod. It doesn't make sense to drop 40 hours of podcasts on your poor iPod for a two-hour drive to the cabin, does it? You can control the number and size of podcasts by configuring iPod preferences in iTunes' Preferences (see Figure 11.21).

Windows Media Player

Windows Media Player needs a little help getting podcasts down to your portable audio devices. It's not terribly difficult, just needs some setup to get things going. Briefly, podcasts are downloaded by a podcatcher and then Windows Media swings into action, using auto playlists to load and sync the podcasts.

To automate the download and sync of podcasts to your portable media device:

1. Install and configure a podcatcher to deliver podcasts to a folder on your system.

2. Configure the podcast folder as a Monitor folder in Windows Media Player. As podcasts are downloaded they will be deposited in this folder.

Open Windows Media Player. Downloaded podcasts will be assigned playlists automatically. These playlists can be played, synced to portable media devices, or burned to CD to play in the car. Windows Media Player also has the ability to automatically sync a playlist when a portable device is connected (see Figure 11.22). Using this feature, you can automatically sync the podcasts to your media device.

Figure 11.21
Configuring iTunes Preferences to only download the most recent podcasts will avoid killing your iPod.

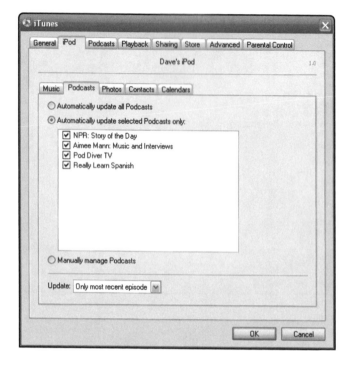

Figure 11.22
Use device properties to configure Windows Media Player for automatic synchronization.

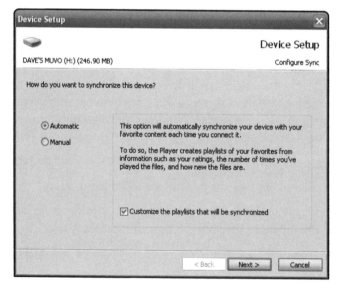

To configure your portable media device to sync automatically:

1. On the Tools menu select Options to open the Options configuration window.

2. Choose Devices to display the devices currently connected to your computer (see Figure 11.23).

Figure 11.23
Media Player displays a list of connected devices.

3. Select the device you want to configure to autosync and click Properties.

4. In the Devices properties dialog, check the checkbox next to Start sync when the device connects.

5. Click Settings to manage the playlists that will be synced automatically.

6. A wizard will run the first time you configure autosync. Choose Automatic to configure the automatic sync function. Click Next.

7. Choose the playlist belonging to the podcast you are configuring and click OK.

The next time you connect your portable media device, the podcast will automatically sync to it.

Windows ActiveSync
In addition to the ways Windows Media Player can sync podcasts, you can also use Microsoft ActiveSync to sync podcasts to Windows Mobile-based PDAs. By designating a sync folder in ActiveSync settings, you enable it to copy any files detected there to the PDA for play offline. The podcast files should be able to play in your PDA without modification or transcoding.

Winamp

Unlike iTunes and Windows Media Player, Winamp comes with no inherent ability to autosync to a portable media device. Don't let that keep you from using it for podcasts, though. With the simple addition of the manual step of copying your media to the portable media device, you're done and ready to go.

Winamp will automatically index the podcasts in your watch folders, adding them to your Music Library. You can play them, burn them to CD, or copy them to your portable media device for offline listening.

HAVING FUN WITH PODCASTS

What I'm Listening To

Title: PDTV 2.2 (Video Podcast)
Artist: Joseph Cocozza
Source: Pod Diver TV
Player: iTunes

Well, you've seen about a thousand options for gathering, consuming, and digesting podcasts. How about we just have some fun with them? What kind of podcasts are there? Is there anything I would like?

In a word? Yes.

C'mon! Do you really think that out of the over 10,000 podcasts out there you won't find at least one you'd like? What's your passion? Do you like music? 2,440 podcasts! Information Technology? 1,446! Business? 374. Big wave surfing? Google lists 253 hits for "big wave surfing."

So, read on. I'll show you where to find the latest on your daytime dramas, director's commentaries on *Battlestar Galactica*, news, weather, even-gasp–podcasts on podcasting!

Follow Your Favorite Shows

Producers of television shows and the celebrities who act in them have found podcasting. Well, actually, for the most part their publicists have. You can get the latest dish on your favorite celebrity, insight into the inner workings of a television series, and updates on your favorite daytime drama.

Look for podcast feeds where you get other information on these shows. Search iTunes or Podcast Alley. Ask the Podcast Pickle. They're out there and, like I said earlier, they want you to find them.

Directors' and Producers' Commentaries

Many directors and producers love to talk about their motivations for the way they shot certain scenes or why they chose certain scripts. They talk about character development, plot twists, basically anything they want. One of the more interesting podcasts by a director is the *Battlestar Galactica* podcasts done by Ronald D. Moore, the executive producer of Battlestar Galactica, a series on Sci-Fi Television. In this podcast he synchronizes the commentary with events unfolding on the screen.

This type of interaction with the audience has earned *Battlestar* an audience that is so dedicated that they actually convinced Sci-Fi to renew the series when they had intended to drop it. Other Sci-Fi series are helping lead the way in podcasting, notably *Stargate Atlantis* (also from Sci-Fi Channel) and *Star Trek Enterprise*.

Daytime Drama Updates

Want to catch up on the latest episode of your favorite soap but have to be out of town? Download the podcast! *Guiding Light*, a CBS daytime drama, publishes a daily podcast to let fans keep up with things when they can't be there to watch.

Other shows are adopting this theme as well. NBC podcasts portions of the *Today Show*, ABC offers *Good Morning America* and *Nightline*, and all networks are producing special podcasts to support regular shows and television specials.

The Dish on Your Favorite Celebs

From podcasts done by your favorite celebrity to tribute podcasts done by their biggest fan, news and dish about celebrities is just an autosync away. Try the Hollywood Podcast for a behind the scenes look at Hollywood from insider Tim Coyne. AOL Moviefone also has a weekly podcast featuring the celebrities who are starring in newly released movies. Look for the entertainment or movies and file categories of your favorite podcast directory.

Fan Sites

Many fans of television shows blog and podcast about their likes and dislikes. You'll hear their take on various episodes, what they like about the stars, and occasional tidbits about their own personal philosophies. It is strangely compelling to listen in on a show that no one else has probably ever heard, maybe no one else will ever hear. You're sharing a few moments with another individual you'll never see, never really know, but they are talking to you, and you just can't quite turn away…

News, Weather, and Sports

News agencies have gone for podcasting in a big way. From podcasts of individual news stories to podcasts of the nightly news, you can find literally gigabytes of news and information on the major news sites. Some are organized into feeds you can subscribe to; others will have to be grabbed manually.

Radio News Podcasts

Radio news is an excellent partner to podcasting. A quick search for news radio podcasts will net several podcasts of radio news. Other sources such as ESPNradio and Virgin Digital Radio can be found here, too. Chances are any talk radio show you like will have a podcast version to grab when you've missed the live show. Just check out its website.

Local Media Podcasts

Some local television and radio stations have begun offering podcasts of their regularly scheduled programming. Visit your local station's website to see if they offer a podcast version of shows and news articles.

Indie Radio in Your Pocket

Independent media, whether it be radio, television, or podcast is a non-commercial real-life exercise in communication. Many podcasts are done as a hobby; some hope to one day make enough to pay the bills. Some want you to hear the songs they like; others just want to talk. Try them out, and subscribe to the ones you like. E-mail the hosts or participate in their forums to let them know you are listening. This encouragement keeps the hosts going at 4 AM when they're still editing their next podcast and have to be at work at 9.

Here are a few good jumping off points to independent music podcasts:

- The Association of Music Podcasting (musicpodcasting.org)
- The Podsafe Music Network (music.podshow.com)
- Podsafe Audio (podsafeaudio.com)

What Is Podsafe Music?

Podsafe music is music created under any license that makes it free to use for a podcast. The artist makes the music available to the Internet community in hopes the songs will catch on and raise awareness of the artist's work. The music may require a license when broadcast on the radio or sold in stores, but is provided for free to the Internet community for podcasting. Podsafe music directories exist for the purpose of helping podcasters locate good music to use in their shows.

Corporate News and Business Podcasts

For those whose incomes depend on being the first to know about new products and trends, the PR Newswire offers an extensive list of podcast feeds (see Figure 11.24) for the distribution of public announcements regarding business initiatives and new product announcements.

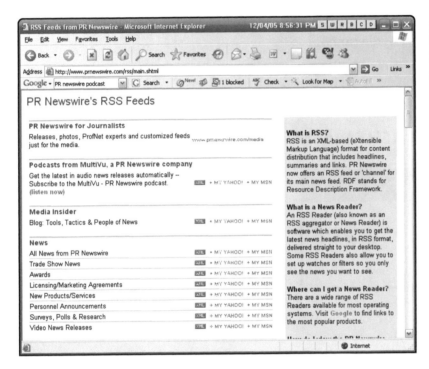

Figure 11.24
PR Newswire is a clearing house for corporate news releases.

Business news and information is another big new area for podcasting. Business moves very fast and many individuals don't have the hour or so needed to sit in front of a TV or listen to a webcast. Podcasting lets them download a podcast to take on the train or plane to keep abreast of important trends in their industry. Some venture capitalists are now using podcasts to subscribe to elevator pitch feeds, podcasts produced by those who want to start a new business but need capital. On a train ride, a VC specialist can hear numerous elevator pitches and perhaps have the next big thing by the time they walk into the office.

Tech News Podcasts

Many podcast shows exist to let geeks like us know about new trends in computers and technology. These technical news shows feature new products and computer systems, teach listeners how to perform certain feats of technical wizardry, or merely report on technological trends. One of the longest running shows of this type is the *Chris Pirillo Show*. Chris, the founder of the

Lockergnome technology enthusiast's website, broadcasts daily from Seattle, WA, on streaming Internet radio. Archives of the show are available as podcasts by subscribing to one of the podcast feeds listed on the chrispirilloshow.com home page.

Another great tech news podcast is produced by CNet's news.com. This daily tech news podcast covers technology from a magazine-like perspective—contributors to stories on other parts of the news.com site contribute to the podcast and are invited to discuss their stories in depth.

Other good tech news podcasts directories:

- Tech Podcasts on ZDNet (podcasts.zdnet.com)
- The tech section of your favorite podcast directory

AS IF PODCASTING ITSELF WASN'T NEW ENOUGH...

What I'm Listening To

Title: I Go Crazy
Artist: Paul Davis
Source: MSN Music
Player: Creative Nomad MuVo NX

Podcasting is literally creating a revolution in the way people communicate. The ability to acquire and then travel with digital media has opened up a host of possibilities. The release of the video iPod; video players from Creative, iriver, and Samsung; and Windows Mobile PDAs with video capability all have led to the next major innovation:

Video Podcasts

As if some of the folks doing podcasts weren't scary enough as a disembodied voice, now we can see them! Amaze your friends when you whip out your new video iPod and show them something other than the latest music video (not that that's not cool). Show them video of a podcast host picking his nose while discussing the merits of autonomous air refueling of the Beechcraft King Air C-12 aircraft. You're going to be the life of the party!

All seriousness aside, video podcasting will bring the ability to create interesting new productions. It will take a while for podcasters to ramp up to it, but I fully expect some great new ideas (see Figure 11.25) to spring from this capability (once we get tired of seeing each other naked).

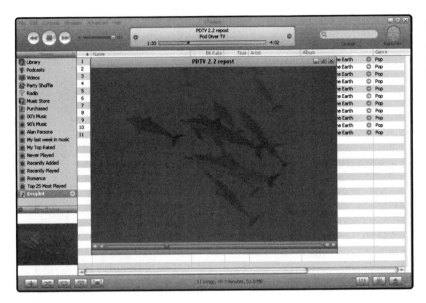

Figure 11.25
A video podcast from the Pod Diver TV podcast feed.

Some good video podcast directories:

- Pod Guide.TV (podguide.tv)
- Podcasting News' Video Podcast Directory (podcastingnews.com)

Education Podcasts

Podcasts are a great way for teachers and students to apply learning principles using Internet technologies. Imagine being able to take your homework from Spanish class in the car or on a trip. Students can also use podcasting technologies to create reports and turn in assignments in distance learning programs. Teachers can get continuing education from anywhere in the world, making it easier for isolated educators to keep abreast of current developments in their area of study.

Podcast directories for educators are starting to appear (see Figure 11.26), making it easier to locate podcasts (or podlessons) in specific areas of interest.

Some good educational podcast resources:

- The Education Podcast Network (epnweb.org)
- Stanford on iTunes (itunes.stanford.edu)
- Teach 42—Education and technology (teach42.com)

Figure 11.26
Educators can find pod-
casts using EPNweb's
podcast listing.

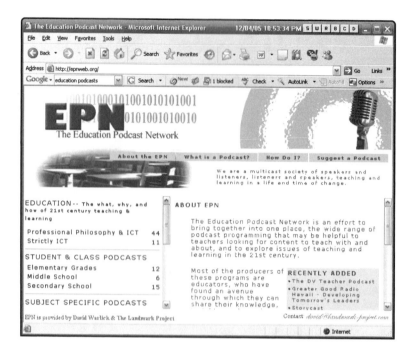

Podcasts in the Car

Most localities frown on the use of headsets while driving in a car. Even if you don't have a car
adapter you can still enjoy your favorite podcasts in the car, with one quick change.

Burn them to CD. Having the podcast on CD allows you to slip it into your CD player and just
drive, listening to the podcast without headphones and without worry of losing your portable
media device or having it stolen.

As podcasts are downloaded into your media library, they are added to playlists in Windows
Media Player and Winamp. iTunes doesn't do this automatically, but can be helped along by
creating a playlist manually. Once the playlist is complete, it is a simple matter to use it to burn
the podcasts to disc. This is especially useful if the podcasts are something you would like to
keep in a permanent archive. An example of this would be the Really Learn Spanish podcasts
produced by Johan van Rooyan. This series of podcasts helps you learn (or relearn) Spanish and
can be used in the car while you drive to work to help you prepare for that trip to Spain. Having
an archived resource like this on disc is also handy when you want to refer to an older episode
that may no longer be stored in your podcast download folder.

The process of burning CDs was covered at length in Chapter 10. If you need a quick refresher
(or dropped in at this chapter) feel free to refresh your memory about the topic. We'll be waiting
for you in Chapter 12.

NEXT UP...

In the next chapter I'll give you some tips on how to create your own podcasts or streaming media broadcasts. Using free or low-cost technology and applications it is possible to create very high quality podcasts to share with friends, business contacts, or the general public. Some enterprising folks are using podcasting to work their way through college or even to make a living. You'll learn how to select equipment, create and upload podcasts, and list your podcasts in places where they will be more likely to attract attention. Join us in Chapter 12 to learn how to be your own radio station!

the gadget geek's guide to

Portable Media Devices

12

Start Your Own Radio Station

You've seen podcasting from the consumer end; now, let's take a look at what goes into creating a podcast. In this chapter I'll show you the equipment and software you'll need to produce your own podcast. You'll learn the steps required to record, edit, and publish your own podcasts. I'll show you how to list your podcast in the podcast directories and how to promote it. I'll also introduce good old-fashioned Internet broadcasting using on-demand and streaming media.

BEGINNER'S PODCASTING

What I'm Listening To

Title: Volcano
Artist: Damien Rice
Source: Purchased from MSN Music
Player: Windows Media Player 10

Your first podcast may seem very complicated and labor intensive. Once you've done a few, though, it will become automatic. You'll know which recording levels work best, how to position your mic to get the best pickup, and how to fill in the minutes with interesting conversation. You'll start planning to improve your productions, adding interviews and musical intros. You might even discover that you have a place among the ranks of podcasting's elite.

In this section I'll give you a tour of the basics of podcasting, show you how to set up a basic podcasting system, and how to record your first podcast. I'll concentrate on free or low-cost packaged podcasting systems, but will introduce you to the more advanced recorders and mixers to give you an idea about where you can go as your skills improve.

Getting Started

Before you start gathering microphones and bumper music, spend some time deciding what your show is going to be about. Subscribe to a few top ten podcasts and see what they have in common. You'll see that they are not necessarily professionally recorded, may not even use music to open and close the show, and might not even be focused on a certain topic. Try to identify what people find compelling about these shows. Identify your own hooks and themes. Write a few sample scripts and mull them over. When you've found your voice, get your system set and start recording.

Podcasting Software

The first podcasters had to know how to record and mix audio. They had to transcode their podcasts into MP3 from WAV or AIFF after they were produced, and they had to figure out how to write the RSS syndication file to put things up on their blog. They spent long evenings getting the sound right, manually editing the RSS podcast feed, and uploading it all to their blog server.

Things have really changed in just a few short years. Several applications now exist to help you record and edit your podcast. Decent microphones are available to help you get the best voice quality for your show and podcast sites are set up to help you create your podcast feed and load in your shows.

Windows Podcasting Software

Windows users have a choice of several good podcasting tools. I'll discuss a few in this section. Others are listed in any of several podcasting software directories such as podcastingnews.com (look for the Podcasting Software link on their home page).

Many of these applications will let you produce and upload your entire podcast. Others may simplify either the recording process or the upload process. At least one is designed for a specific podcast hosting site (BlogMatrix), while others can support a number of hosts. All you really need is a good microphone and you're all set!

Audacity

Audacity (see Figure 12.1) is a sophisticated digital media recorder and editor for Windows, Linux, and Macintosh computers. It provides a graphical means of recording, editing, and mixing audio files. Available free from audacity.sourceforge.net, it is easy to use and allows you to record and mix multiple tracks as you create your production.

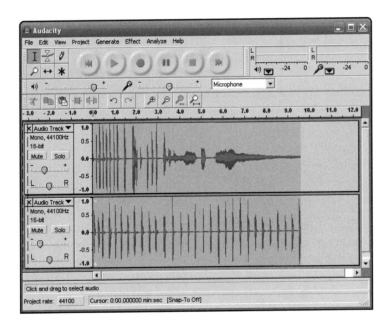

Figure 12.1
Audacity is a free recording and editing package for Windows and Macintosh.

With a tool such as this, one can create the entire MP3 version of your podcast. After creation, the only thing left would be to publish and promote your creation.

BlogMatrix Sparks!

BlogMatrix is a blog and podcast hosting site. Its application, BlogMatrix Sparks! (see Figure 12.2) offers Windows and Macintosh users an end-to-end solution for podcast creation and publishing. It also includes tools for recording Internet radio feeds, podcatching, and blog viewing. You might find yourself spending an inordinate amount of time in this program if you're not careful. BlogMatrix Sparks! is free, but hosting plans on BlogMatrix start at about $5 per month.

Figure 12.2
BlogMatrix Sparks! offers end-to-end podcast creation and publishing.

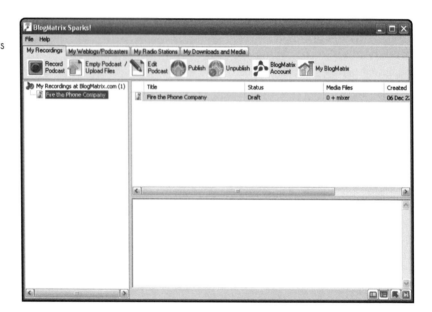

Propaganda

Propaganda (see Figure 12.3) is another Windows podcasting software package. As another application that is linked to a podcasting site (libsyn.com), Propaganda also records, mixes, and uploads your podcast. Propaganda costs about $50 US, but comes with a three month free hosting offer from libsyn.com to help get your show off the ground.

Propaganda uses a powerful timeline interface to let you visually create your podcast by placing audio segments on the timeline in the appropriate position. Volume levels can be adjusted to fade each segment in or out depending on your preferences. The entire composition can be played on your system to test the sound levels and can be saved to an MP3 file or uploaded

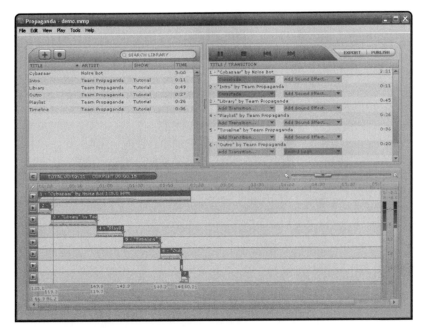

Figure 12.3
Propaganda uses time-lines to visually mix podcasts.

directly to libsyn.com. If you choose not to use libsyn.com, you can use Propaganda's FTP feature to upload to any podcast site that accepts FTP uploads.

Macintosh Podcasting Software

Being the original home of the podcast, the Macintosh computer system has its own share of podcast production packages. The range is similar, with some offering the entire package and others performing part of the process. The podcastingnews.com software directory mentioned above also includes software listings for the Mac.

Garageband

Garageband is a free application bundled with Macintosh OS X. It is designed to let amateur performers develop their own musical compositions. It can be used to record and mix your podcast, especially if you plan to create your own music. Features include the ability to create and use musical loops and drum tracks, record MIDI tracks, and mix all of these with recorded vocals.

Audacity

As previously mentioned, Audacity is an excellent tool for recording and mixing your podcast on a Windows, Linux, or Macintosh computer. As far as podcasting is concerned, it has superior features to Garageband in that it can support higher sampling rates for higher quality recordings and can also manage pitch and noise removal in the editing process.

M-Audio Podcast Factory

M-Audio produces the Podcast Factory (see Figure 12.4), a hardware/software solution that provides the ability to record, mix, and publish podcasts on the Mac. At about $180 US, it is one of the few solutions for the Mac that does not rely on several separate tools to perform the podcast production process.

Figure 12.4
Podcast Factory includes everything you need to create a podcast.

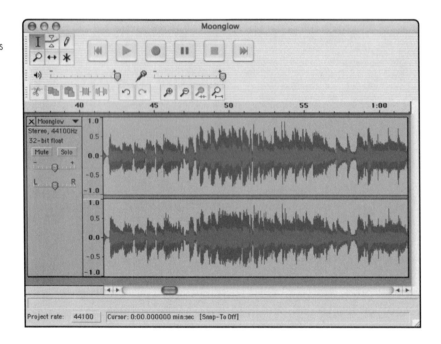

Feeder

Feeder is a solution by Reinvented Software of the UK that is designed to publish podcasts. This application can create iTunes tags for your podcasts to present them in the best possible light in the iTunes podcast directory.

Feeder's specialty is the listing of the podcast in the podcast directory. The podcasts themselves have to be created in another application (such as Audacity).

Podcasting Equipment

As you prepare for recording a podcast, you'll make equipment selections based on quality and price. In this section I'll show the different types of equipment that can be used in podcast production and describe the major differences among various types of equipment.

Microphones

A quality microphone is critical to a quality podcast. Podcasting follows the age-old GI-GO (garbage in-garbage out) rule of computing. Put poor quality voice into a podcast and there is no amount of software that will fix it.

In this section I will describe the different types of microphones you can choose for your podcast rig. There are certainly differences and you'll make a better choice when you are aware of all the options.

Dynamic Mics

Dynamic microphones (see Figure 12.5) use sound waves to generate a signal in a magnetic coil. The signal is transmitted to the recording system as a pure analog signal. The sound recorded may be warmer and more lifelike, but dynamic mics do not boast the same sensitivity levels that can be found in condenser mics (discussed next). Dynamic mics come in a variety of designs and are favored in conditions where background noise is an issue. Dynamic mics are excellent for voice recording and for recording instruments such as acoustic guitars and drums. They have applications in radio broadcasting and can be an excellent choice for a podcast where spoken word or live music is the most prominent feature.

Figure 12.5
The Electro-Voice RE20 dynamic studio mic is used by many professional broadcasters.

Condenser Mics

Although they might look very similar (see Figure 12.6), condenser microphones are more sensitive than dynamic microphones. They can be so sensitive that care must be taken to prevent the incorporation of outside noise into the recording. Condenser mics can be purchased with various recording patterns and filters to ensure the highest quality recording possible. They support a wider range of audio frequencies and are used for high-fidelity recording of music and sound effects.

Figure 12.6
Visually indistinguishable from a dynamic microphone, the condenser mic (right) offers better frequency response.

Portable Mics

Some mics are designed to travel to interviews for the "man on the street" approach to your podcasting. They are ruggedly built and can take a lot of abuse (see Figure 12.7). Along with a digital recorder, they form the kit necessary to record your show outside the studio. If you plan to do interviews or take your show to outdoor locations, you will want one of these in your collection.

Figure 12.7
The Electro-Voice 635A mic is a staple of radio and television news reporting.

Studio Mics

Studio mics (see Figure 12.8) are designed for the highest sound quality. Many are mounted on stands that insulate them from any vibration that would be picked up in the recording. They can be condenser or dynamic microphones and are typically designed for highest sensitivity. This makes them unsuitable for outdoor or portable use. If you do a lot of speaking or record music for your show, you'll want one of these to get the highest quality recording possible.

Figure 12.8
Studio mics can get pretty elaborate.

Lapel Mics

Commonly referred to as "lavaliere" mics, the lapel mic is typically used by those conducting live television shows or performing in front of an audience where the microphone should be hidden or as unobtrusive as possible. Unless you plan to start video podcasting, you won't need one of these right away.

Pop Filters

Do you pop your Ps? Perhaps you should pick a perfect pop filter to prevent the production of popping P sounds!

Pop filters block the rush of air produced with sounds such as P, B, and T. They can also control sibilance produced with the letters S and Z. They allow the sounds to pass through to the mic, but prevent the rush of air that overdrives the mic's recording elements.

Pop filters can be built into the microphone or can be added between the mic and the speaker (see Figure 12.9). When you read the specifications of the microphone you are evaluating, look for references to pop filters to determine if you will need to buy a separate pop filter.

Figure 12.9
The VAC pop filters from popfilter.com are used by many recording and voiceover artists.

Microphone Directionality Patterns

Microphones are designed to pick up sound in certain patterns depending on their intended use. Obviously you would want a different pattern for a microphone designed to pick up the voice of a single speaker than for one designed to pick up the voices of a choir.

Cardioid

Microphones exhibiting cardioid directionality are more sensitive to sounds coming from the front of the microphone. This attribute is desirable when you want to pick up vocals or tones from a musical instrument while eliminating extraneous noise.

Omnidirectional

An omnidirectional mic picks up sound equally from all directions. These mics are great for recording a group of vocalists or an ensemble of musical instruments.

Other directionality designations exist, such as shotgun or figure eight, but are not really germane to our discussion of podcasting. You can Google "microphone directionality" for more information on this topic.

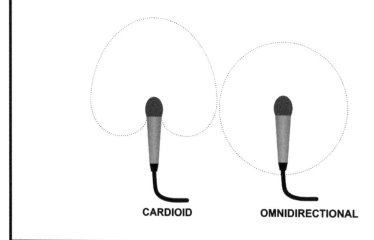

CARDIOID   OMNIDIRECTIONAL

Editing Gear

Once you have the means to record the sounds you make, you'll want some way to actually record them. In this section I'll take a look at digital recorders and mixers. Recorders let you record interviews away from your computer to bring back for your show. Once the recordings are back, use a mixer to combine them with music, sound effects, and other voices for a complete production.

Some podcast applications allow you to accomplish mixing on your computer. This is fine for most uses. Sometimes, however, you'll want to mix live when you have multiple hosts or in-studio interviews. A mixer takes multiple inputs from microphones, instruments, or playback equipment and mixes them in real time.

Recorders

Digital recorders allow you to record your show even when you aren't in your studio. This is helpful when you are covering events that are not occurring in your bedroom (or wherever you usually podcast from). A good digital recorder should cost about $80, with professional models (see Figure 12.10) running in the $500–$700 range. Some professional models offer limited editing on the recorder, helping you split the recording process into multiple files to be used as individual tracks in your editor.

Figure 12.10
The Marantz DM670 professional recorder includes limited editing capabilities.

Mixers

When you plan to include two or more inputs (such as microphones from two speakers) in your podcast, you'll need to mix them as they are recorded. Using a mixer (see Figure 12.11) allows you to mix one or more microphones with input from an instrument or playback from a recording device. This simplifies the process of final mixing on the computer because it is much simpler to record multiple sources live than to mix them and keep them in sync afterward.

Figure 12.11
The Soundcraft Compact 4 mixer sells for about $100 US and can mix four inputs.

Creating a Podcast

The process of creating a podcast is as unique to each individual as his own fingerprints. You'll use different microphone placement, different input levels on your mixer, different mixing software, and different editing software. Your topics will differ and you will invite different guests. Some of your podcasts will be acclaimed as original, witty, and insightful (common critic words). Others will not be so well received. The important thing is that you have your voice and you can send it out there to see who responds.

In this section I'll outline the basic steps to creating a podcast, providing examples along the way.

Writing a Script

Without a script—or at least an outline—your podcast will be in danger of long quiet stretches while you think of something witty to say and may segue into off-topic issues. A script instills a discipline to your work and keeps you firmly focused on your objectives.

Create an Outline

Start with a broad outline of what you hope to cover in your show. Use major topics in level 1 headings of your outline and add minor topics related to these major themes as level 2 subheadings. If you have specific discussion points you want to make, insert them as level 3 lines within the appropriate level 2 subheadings.

An example might look like this:

DAVE'S RAVES EPISODE 1

Intro

Love this weather!

    Temperature

        Too cold for some

        Perfect for wetsuit surfing

    Winds

        Direction

        Wind effects on wave formation…

Notice that the outline suggests some discussion points in each section of the show to keep conversation flowing. As you actually record the show you will decide whether time permits you to use these suggestions. Having them there will keep you talking and keep your audience interested.

Fill In Topics

As you complete your outline you can go back and add additional information to each heading such as facts and figures related to each discussion point. If you have statistics about wave heights for a given wind speed, put them in! When you get to that discussion, you can appear wise as you respond to your guest's comments on the prevailing wind with, "Well, Steve, they do say that wave heights of up to 14 feet are not all that uncommon at these wind speeds. Sounds like a good day to wrap this podcast and go hang ten!"

Recording the Show

When you begin recording the show you'll follow your script to keep things on course. As your show progresses you'll also want to improvise a little to keep things loose. Your audience will be able to tell if you are too stiff and scripted. With practice you'll glance at the script from time to time to be sure you left nothing important out, but you might go off into (planned) topics for minutes at a time without referring to the script.

Following the Script

The script is a framework. Nothing more, nothing less. If you follow it too closely the show will be stiff and over all too soon. Feel free to wander within each topic, fleshing it out and exploring it fully. Glance at the timeline and your script to know when you need to move on, but feel free to let the show flow.

Improvising

Occasionally you'll get to the end of the script and still have five minutes left. This is when you'll find out how good you are at improvising. Open a natural conversation with your guest, or just spend some time talking about the next episode. If you completely run out of ideas, remember that you are not actually live. Just cut the mic and stop the recorder. Record a few segments to fill in certain topics and mix them into the show at the applicable point.

Do-Overs

If you completely blow a segment, just redo it. You can use your editing software to splice the new section into the show. A good recording rig can divide your show into segments or clips that make this editing easier. If you don't have a collection of clips, you can also just cut and paste the sections at the appropriate points.

Interviews

Interviews are a great way to fill the time. If you have an Internet softphone such as Skype or ineen, you can actually record telephone interviews with your editing program. Mix them with intro music and other voice segments to make a complete show. In-person interviews can be recorded through a mixer or, if you only have one mic, by getting cozy with your guest and sharing the mic.

Plan your interview questions and include them in your script. This will help you remember that cool topic you wanted your guest to talk about. Remember that you invited your guest to speak. Sometimes it is easy to talk over your guest and not give him a chance to finish his thoughts; this is not only rude, but also is likely to cut your show short when your microphone is tossed into your fish tank.

Completing the Podcast

Once all the segments of your podcast are recorded, you can begin the process of editing. You will cut each segment to size, removing any dead air from before and after the actual talking. You'll then splice them, adding bumper music to ease the transitions between segments. If you are lucky enough to get a sponsor, you'll also be adding their messages between segments like your very own mini radio commercials.

Cutting and Splicing (Editing)

Cutting and splicing segments will become easy with practice. At first you'll spend a lot of time finding the beginning of each segment and trimming it. This gets easier as you become more familiar with your editing program. If you are using a complete podcasting package, you'll record, edit, and publish all in one program (see Figure 12.12). Take the time to completely read any documentation available on your program to get the most from all its features.

Figure 12.12

Propaganda simplifies the process of editing your show.

Tip

If you plan to use bumper music (the music used to intro a show), be sure the music you select is "podsafe." Podsafe music is free to use for podcasting and does not require you to pay a licensing fee to use it in your podcasts. Google "podsafe music" for directories and links to podsafe music sites.

Saving the Results

When your show is completely edited you'll save it. Most editors can save the components of the show (see Figure 12.13), keeping mixing information available in case you want to do further editing. It is important to save your show in this format. If you do, you can come back and make minor changes to correct mistakes or add information you inadvertently left out.

Figure 12.13

Save a complete copy of your podcast to enable the ability to make changes later on.

Most podcasting tools also let you save to MP3, WMA, or even AAC. Even if you are publishing directly online, it is wise to save your podcast locally before sending it up. In this way you have a backup of your podcast in case something goes wrong.

Publishing a Podcast

Once your podcast is complete you'll put it online. Hopefully you settled on a host before your first episode was recorded. (You have, haven't you?)

In this section I'll discuss hosting options and give you some tips on selecting the host best suited to your needs.

Finding a Host

Podcast hosts are not too difficult to find. Most are blog sites that have expanded their offerings. If you listen to a lot of podcasts, chances are you already know about a few providers. If there's one thing podcasters like to talk about it's their hosting experiences. Good or bad, you'll hear about them. Listen to their accounts of good experiences and Google those hosting sites. Search the podcast directories for links to hosting providers. If you settle on BlogMatrix Sparks! or Propaganda for your editor, you can even use the provider integrated into their products.

Blog Hosting Sites

Major blog hosting sites often host podcasts as well. Sites like BlogMatrix also offer the ability to include podcasts with your service. If you already blog, check out your host's offerings in this regard. Even if they don't host podcasts directly you can often just upload the MP3 file and link to it with a directory such as iTunes. The feed file gets a little trickier, but can be done (the first podcasters did it!).

Dedicated Podcast Sites

If you are using a podcast application that includes the ability to upload to a certain site, chances are it is a site in this category. Libsyn can accept podcasts that are uploaded directly. Libsyn is the host of choice for the Propaganda podcast editor.

When setting up your editor to upload to a dedicated site, you'll need to create an account. This is often done within the editor, and may require a credit card to manage ongoing subscription services for the site.

Hosting Your Own Podcast

If you have a website, you can host your own show. All you need to do is upload the podcast media files and an RSS syndication file. The RSS file lets subscribers be notified of updates in your podcast. When you add episodes, you'll update this file to include information about the new episode, such as URL for download, a description, and other information such as a brief synopsis of the contents of the podcast.

It is helpful to use blogging software to maintain your podcast site. This lets you use blog entries to define the different podcast episodes. Simply write a short intro, include a direct link to the

MP3 file for those reading your blog, and edit your RSS podcast feed to notify subscribers that there is a new podcast.

Getting the Word Out

The RSS podcast feed file is the most important aspect of ensuring listeners are updated whenever a new episode is uploaded. This file is created automatically by some upload tools and can be created by applications such as Feeder. Once the feed is uploaded, the address of the RSS-formatted feed file is the address you'll use to list your podcast with podcast directories.

Listing with Podcast Directories

Podcast directories index thousands of podcasts and simplify the process of locating a podcast that specializes in a topic you find interesting. To attract listeners, you'll list your podcast and use descriptive terms that will maximize your chances of being in the search results when a directory user is looking for information contained in your podcasts. The podcast title and description can get you quite a few hits. Your RSS file may also be indexed so topics of individual episodes will be available for searching as well.

Word of Mouth Advertising

As your base of loyal listeners increases, they will share your show with friends. This word of mouth advertising is a great tool, but depends on your ability to hold and entertain your audience.

Encourage listeners to tell their friends. Offer an incentive, such as a drawing for subscribers who are named as referrers by those subscribing to your feed. Of course this means you'll have to add a subscription form to your site...

Linksharing

Sharing podcast links with other podcasters is a great way to promote your podcast. They get traffic from listeners of your podcast, and you benefit from their subscriber base as well. The more links you can share, the better your chances are that Google or another Internet search engine will list you when users search for key terms in your podcast title or description.

Shameless Plugs

Don't be above blogging about your podcast and mentioning it whenever you are in online forums. By providing links in all these ways you are increasing the opportunities folks have to find your podcast. This profusion of links will also raise your ranking with Internet search engines. Put your podcast feed URL in your e-mail signature and send lots of e-mails and newsgroup postings. Find and comment in forums that are appropriate to your area of expertise. They will appreciate your participation and you'll get more feed subscriptions.

Podcasting Organizations

As your experience with podcasting grows you will want to work together with other podcasters to raise your subscriber base and promote your podcast. Many organizations have been formed to promote podcasting and podcasting techniques. You will find articles posted on their websites giving you tips and tricks that will make your own podcasts better. You'll also be able to participate in online discussions, local meetups, and trade shows sponsored by some of these groups.

Association of Music Podcasting

The Association of Music Podcasting is an organization of podcasters who create podcasts featuring podsafe music created by independent artists for use in podcasts. Their website (musicpodcasting.org) features a library of podsafe music and a directory of member podcasts.

TechPodcasts.com

TechPodcasts.com is a group for podcasters specializing in technical podcasts. Members must have podcast at least five shows and meet certain criteria for website design and usability. Members participate in roundtable discussions and are listed in a directory on the TechPodcasts.com website.

Portable Media Expo

The Portable Media Expo has become the premier trade show and expo for podcasters. A trip to the show lets you hobnob with the industry's giants. You'll see podcasts being recorded live and you'll get up to the minute information about the newest tools and techniques for podcasting. The Annual People's Choice Podcast Awards promoted by Podcast Connect, a separate organization, are held in conjunction with this annual convention and expo.

BROADCASTING

What I'm Listening To

Title: Witness
Artist: Sarah McLachlan
Source: Surfacing (Ripped from CD)
Player: Windows Media Player 10

Lest podcasting steal the show, I should tell you that there are other options for distributing your own personal digital media show. If you aren't a huge fan of the podcast you can still create professional media for use on the Internet.

In this section I will show you how to publish your media creations for download or streaming use. While the former will be nearly indistinguishable from podcasting, the latter will differ quite a bit.

On-Demand Media

Prior to the podcasting explosion, folks uploaded media files to their blogs or put them on their websites. These media files did not differ much from what is commonly seen in today's podcasts. The only significant difference between on-demand media and podcasting is the addition of an RSS podcast feed to alert consumers that a new episode is available.

Long before Adam Curry began using RSS to advertise his audioblog entries, on-demand media was the king. Websites dedicated to digital media have been available for over a decade, serving to distribute photos, videos, and music to millions of web consumers. These sites used a variety of delivery mechanisms including Windows Media, Apple QuickTime, Real Audio and Real Video, even ancient media types such as MIDI and WAV files.

Recording Media

Recording media for broadcast is no different from creating a podcast. Applications such as Audacity are excellent for this purpose. When you create your recording, be sure to consider its intended use. Spoken voice will not require super-high sampling rates, but music would definitely get a boost from high rate recording. Your intended audience will help determine which media codec you will use. If you are targeting a predominantly Windows crowd, use WMA. If you are looking for Mac fans, use AAC. If you don't care who listens, as long as they listen, consider using MP3 audio or MPEG video for greatest compatibility.

Publishing Media

Publishing your on-demand media files is as simple as uploading them to your website. Put a link to the file on your web page and you are ready to go. If you attach the file to a blog article, it will be available to anyone reading your blog.

Streaming Media Broadcasts

Streaming media is another way to get your message out. This real-time service requires some special tools to produce, but is great if you want to offer a live show to your listeners.

Windows Media Encoder

Microsoft makes the Windows Media Encoder available free of charge to those desiring to create Windows Media content. It can be used to record Windows Media audio (see Figure 12.14) and video files or to distribute audio or video live to Windows Media Player.

Figure 12.14

You can create your own WMA and WMV files with Windows Media Encoder.

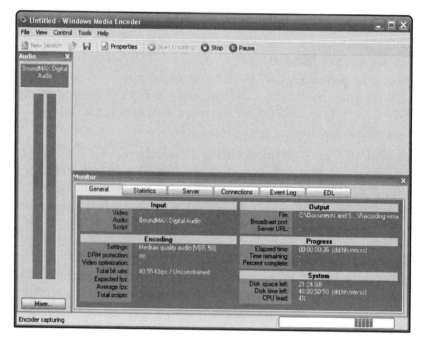

Creation of a media file is fairly straightforward. Windows Media Encoder offers wizards to guide you through the process of configuring the recording settings and output type. For live shows it will actually serve as the distribution server for the media stream as well, enabling you to broadcast live on the Internet for free. Windows Media Encoder can also upload content to the Windows Media Services on a Windows Server system for broadcast to larger numbers of listeners, but can definitely manage a few feeds without adding this additional infrastructure.

Apple QuickTime Broadcaster and QuickTime Streaming Server

Apple has its own system for broadcasting live feeds over the Internet. The QuickTime Broadcaster can send media hurtling across cyberspace as well as any other tool can. QuickTime Broadcaster broadcasts in MPEG-4 and can interface with Apple's QuickTime Streaming Server for industrial strength broadcasting.

Webcams

Webcams are another live communication opportunity made possible by digital media. With nothing more than a USB video camera and some software (usually available with the camera), you can show your interior decorating skills to the entire Internet. Webcams are valuable tools for keeping in touch with distant friends and relatives. They are great for video conferencing and are used extensively in video chat applications.

WebJay

One last Internet phenomenon is WebJay, a playlist sharing site (see Figure 12.15). This site allows users to host their own music playlists of content they legally own. Other users can browse these playlists and are able to play the songs directly off the publisher's system. This "sharing" does not violate any current copyright laws because users are not sharing songs that are restricted. WebJay removes any links discovered to be using protected songs. This tool is great for new artists to publish their music and get playtime.

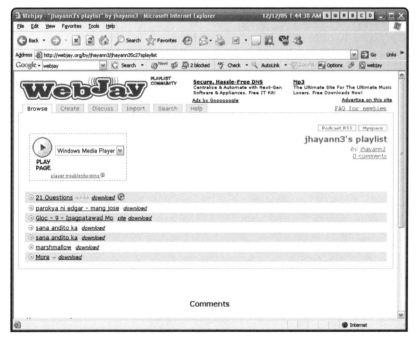

Figure 12.15
WebJay allows users to share playlists containing links to freely distributed songs.

You can create a Windows Media playlist of songs that you have published on the Internet or of songs that you have linked to. After you upload this playlist to WebJay, other users can download it and enjoy the songs you have linked to. As more folks link to your songs, the links form a sort of index. By clicking on a song you see everyone who has published a playlist including the same song. Clicking on their playlists leads you to other songs they liked and that you might possibly enjoy.

ALMOST THERE...

Well, there's not a lot more to teach you about managing your portable media and portable media devices. About all that is left is to have some fun with a glimpse into the kind of cool things we'll be doing with our media. You'll see the chapters come together and how to use what you've learned to get the most from your media devices. I'll also do a fun bit about a day in the life of a gadget geek. It will be fun, and a little corny, so bear with me.

Portable Media Devices

13

More Than Media

This is the chapter where we bring all the tips and tricks together. I'll show you how to use the additional features and capabilities of your portable media device. In addition to being able to store music and video, many portable media players can act as portable disk drives. They can also serve as voice recorders to enable you to record memos, interviews, and other random sounds. The media players with photo capabilities let you share your photo library with family and friends, and PDAs and mobile phones are beginning to offer media features as well. I'll complete the chapter with a tour of a day in the life of a hardcore gadget geek, bringing it all together to show you what is possible with this technology.

TRANSPORTING DATA WITH YOUR PORTABLE MEDIA DEVICE

What I'm Listening To

Title: C'est La Vie
Artist: Komatsu Ayaka
Source: Bishojo Senshi (Pretty Guardian) Sailor Moon Single
Player: Windows Media Player 10

Many portable media players are able to transport data files as a portable disk drive. With proper preparation, you can connect your media player and copy files to it from your computer. Both iPod and other media players offer this functionality. The process will differ depending on which device you are using.

Mounting Your Portable Media Device as a Disk Drive

Whether connected by direct USB connection or docked in a docking station, your portable media device may have the capability to store and transport files between computer systems. This functionality gives you some of the features of a personal digital assistant without the high cost and limited media functionality.

In this section, I will show you how to connect two common media players to a computer to enable this functionality.

iPods

iPods can be configured with iTunes to present a limited amount of disk space to the computer's operating system as a portable disk drive. This configuration is accomplished in iTunes Preferences (see Figure 13.1).

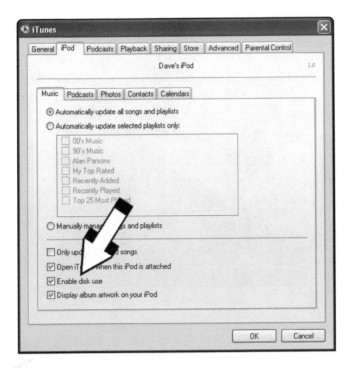

Figure 13.1
iPod's disk drive settings are managed in iTunes Preferences.

To connect your iPod as a portable disk drive:

1. On iTunes Edit menu, select Preferences to open the iTunes Preferences window.

2. Locate and click the iPod tab.

3. Check the checkbox next to Enable Disk Use.

4. The iPod will appear in My Computer (Windows) or on the Desktop and in Finder (Mac).

Meltdown

When using your iPod as a disk drive, be sure to eject it using iTunes before disconnecting it from your computer. Failure to do so can result in disk corruption requiring a recovery operation.

Portable Media Players

Portable media players from iriver and Creative can also be used as portable disk drives. The drive mounting process can usually be accomplished by connecting the portable device to your computer (see Figure 13.2). It will be detected and mounted as an additional disk drive. Any available free space not used for songs will be available for files.

Figure 13.2
The Creative MuVo NX connects directly into a USB port.

Windows Mobile PDA

PDAs running Windows Mobile can be used to transport files as well. Microsoft provides a tool for synchronizing these devices. This tool (ActiveSync) can synchronize more than e-mail and calendar data. You can define folders for synchronization and have them automatically kept in sync whenever you sync your device.

Managing Mixed Songs and Data

If you have songs in your portable media device, you will eventually run into contention for space with your data files. With most media players, this is a simple disk space issue. Watch the free space statistics on your device to see that you still have room for your music and the data you intend to transport.

Determining Free Space

It is important to determine your available free space before copying a lot of data to your portable media device. Running it out of space can corrupt files on the device. The amount of free space available is usually viewed using standard disk management tools, but can be determined in other ways.

Determining iPod Free Space in iTunes

If you plan to transport large files or if you have a lot of music, you will want to monitor the amount of space your iPod has available for music.

To check free space in iTunes:

1. In the Sources pane, click on the icon for your iPod.

2. At the bottom of the iTunes window, view the free space display for your iPod.

> **Note**
>
> The iPod shuffle allows you to reserve free space for data files to prevent taking up all available space with music.

To check free space using the iPod's menus:

1. From the iPod Main menu, use the Click Wheel to navigate to the Settings menu.

2. Select About to display your iPod's statistics, including available space (see Figure 13.3).

Figure 13.3
You can view free space available on your iPod within iTunes.

Other Media Players

Media applications and operating system tools can both be used to determine how much free space is left on your portable media device. File system tools will indicate how much disk space is remaining, while most media players will indicate free space left for music uploads.

Windows Media Player displays free space on the Sync tab (see Figure 13.4).

Figure 13.4

You can view device free space on the Sync tab in Windows Media Player.

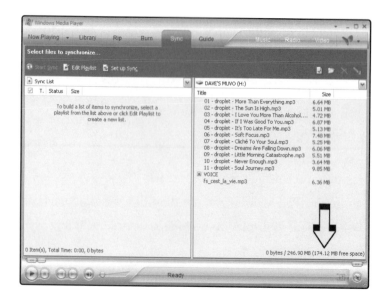

Windows Explorer displays free space in the Properties dialog for the device (see Figure 13.5). Access this dialog by right-clicking the device in Explorer and selecting Properties.

Figure 13.5

Free space can be seen in a device's Properties dialog in Windows Explorer.

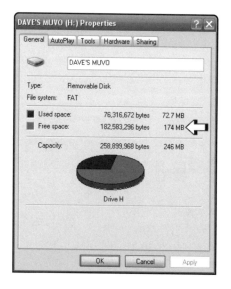

Note

You can also view free space on an iPod using disk properties in Windows Explorer.

BOOT YOUR COMPUTER FROM YOUR PORTABLE DEVICE

What I'm Listening To

Title: Mary Jane's Last Dance
Artist: Tom Petty and The Heart Breakers
Source: Greatest Hits
Player: iTunes

Yes, you can actually boot your computer from your portable media device. There are complete operating systems that you can load into your media device. Your computer can boot from the media device and run applications from it. Access the Internet, recover lost system files, snoop around…

IBM Research has demonstrated the process of booting a computer from an iPod mini. At a show for IBM business partners, IBM representatives loaded a version of the Linux operating system and used it to recover a crashed notebook computer. This is serious geek stuff and waaay beyond what I could hope to cover in these pages, but very good to know about. If this is something that fascinates you, back up your iPod and go looking for the procedure.

Meltdown

Be very careful about any project that involves using your iPod (or other hard disk player) as a boot device. The tiny hard disk inside these players is not designed for computer operating system access and will fail rapidly if accessed in this way for long. Flash memory-based devices do not suffer from this vulnerability (no moving parts), but have limited space and so are not entirely suitable as a boot device.

If you have Macintosh OS X you can actually install it to your iPod and use it as a boot device. Instructions for accomplishing this can be found online. Google "boot from iPod" for detailed instructions. I do not provide instructions here because this can damage your iPod if done improperly (see preceding Meltdown).

VOICE RECORDING

Can't scrape up the funds for a digital recorder for your podcast? Want to record voice memos so you don't forget important ideas for your bestselling novel?

Use your portable media device.

Many media devices are designed for recording voice memos and actually have microphones built in. Just press record and start talking! Other devices need an external microphone or different software to enable them to act as a recorder.

Devices with Built-in Recording Features

Several portable media players offer the ability to record songs and voice memos using a built-in condenser microphone (see Figure 13.6). In fact, the iPod stands alone in its lack of native recording capability.

Figure 13.6
Peer closely at this photo to see the tiny microphone port on the MuVo NX.

While procedures vary from device to device, typically you simply press record and begin talking to record voice messages directly onto your portable media player. Some, such as the Creative MuVo, must be placed into Record mode before recording, but record quality WAV or MP3 files once there. When you next connect the device to your computer, you can copy the new media file to your system for cleanup and editing.

Recording Accessories for the iPod

Some iPods can be used with portable microphone devices such as the Griffin iTalk to record audio. The iTalk enables the voice recording function on these devices to let you record voice memos.

Griffin iTalk

The Griffin iTalk (see Figure 13.7) is an external microphone and speaker that connects to the headphone and serial ports of the full-size iPods (through the 4th Generation). When connected, the iPod recognizes the device and allows the user to make voice recordings. An external microphone can also be connected to a port on the iTalk for better quality recording. This ability has enabled many users to record podcast audio using nothing more than their iPod and this $40 device.

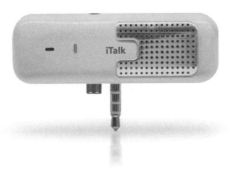

Figure 13.7
The Griffin iTalk enables recording on older, full-size iPods.

The iTalk is controlled with the iPod's Click Wheel and can be used for both recording and playback.

Belkin TuneTalk for iPod

Belkin produces a microphone option with similar capabilities (see Figure 13.8). It lacks the external microphone port, but includes a built-in windscreen to minimize wind noise and breath pops during recording.

Figure 13.8
The Belkin TuneTalk actually looks like a microphone.

Which iPods Can Record?

All full-size iPods have the ability to record when coupled with the appropriate external microphone. The new Video iPod (AKA 5th Generation iPod) does not have an appropriate port for the iTalk or Belkin's Voice Recorder, but can accommodate recordings through the docking port (hardware not yet released).

The 5th Generation iPod can record stereo with the proper microphone setup. Earlier models record only mono audio.

The iPod shuffle, iPod nano, and iPod mini were not granted the ability to record.

Downloading Your Recorded Messages

Once voice messages have been recorded, they can be copied to a computer. With the iPod, this can be accomplished using iTunes. Other media players use the disk drive feature to make the files available to Explorer (Windows) or Finder (Mac).

Transferring Voice Recordings with iTunes

The iPod stores recordings in a playlist called Voice Memos. When you connect your iPod to your computer, iTunes will sync this playlist into the iTunes Music Library. The recording files are WAV audio files and can be copied and used in other media applications. They can also be transcoded into AAC, MP3, or WMA for more compact storage or for use in podcasts.

Transferring Voice Recordings with Explorer or Finder

Portable media players that can be mounted as a disk drive are visible to your computer's file system browser, whether it is Windows Explorer or the Macintosh Finder. Transferring recordings from your media player is as simple as copying them to your system's hard disk.

To copy new voice recordings to your My Music folder:

1. Connect your portable media player and wait for the system to detect it. Close any prompt that opens to suggest possible actions you can take.

2. Open Windows Explorer.

3. Locate and browse your portable media device (see Figure 13.9).

4. Drag and drop or cut and paste your voice recordings to your My Music folder (or any other designated folder).

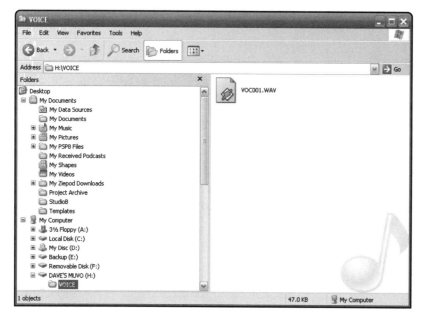

Figure 13.9

Locate your voice recordings on your portable media device.

261

BEYOND MUSIC—PHOTOS AND VIDEO

What I'm Listening To

Title: Whole Lotta Love
Artist: Led Zeppelin
Source: Led Zeppelin II (WMA ripped from
       disc)
Player: Windows Media Player 10

The screen may be tiny, but folks still like to look at your pictures. With the 5th Generation iPod you can display video or a still photo slideshow on your tiny screen or connect to a television to show your pictures and videos full size.

Other devices also offer the ability to store and display photos and video. In this section I'll show you a few of these devices and give you a few tips for their use. I'll also present a few interesting accessories that will help you get photos and video into your portable media device.

Extending Your iPod

The 5th Generation iPod plays video and displays photo slideshows, both on its built-in screen and on a television via a cable connection. This device can store thousands of photos and hours of compressed video. It offers those with no long term data storage space the ability to have their photos and videos nearby. Many users just keep their songs, photos, and videos on their iPod, having no permanent home for these media files on a computer.

Accessories

Several media accessories have been released for the newest iPod. Not having the same connectors as the older generation iPods, some accessories like microphone adapters have been obsoleted by this version, but manufacturers are hurrying to release updated versions of their products.

Connecting Cables

To play video and slideshows on a television, you'll need an adapter cable. The AV Cable from Belkin (see Figure 13.10) connects your iPod to the composite video inputs on a TV or VCR. It will allow you to play video and stereo sound on your home entertainment system.

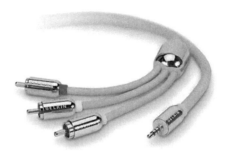

Figure 13.10
Use an adapter cable to play your video and music on your home entertainment system.

Camera Attachments

Camera connectors like the Belkin Digital Camera Link (see Figure 13.11) are now available to allow you to connect your digital camera to your iPod. This capability allows you to use your iPod as a media storage device while on the road. By connecting your iPod to your camera, you can copy your pictures directly to your iPod. With the storage space available on the 30 or 60GB iPod, you can store the pics from an entire vacation.

Figure 13.11
Copy digital photos directly to your iPod Photo or 5th Generation iPod.

Media Readers

Belkin also manufactures the iPod Media Reader (see Figure 13.12). This device will read six of the common forms of flash memory devices used in digital cameras. You can copy photos from these devices onto your iPod.

Figure 13.12
Read six types of flash
memory devices with the
Belkin Media Reader.

Coming Soon

Belkin, Griffin, and even Apple are working feverishly to produce accessories for the newest
iPod. Among the eagerly awaited accessories are microphones for voice recording and remote
control devices (Apple already markets one remote control solution as part of their Universal
Docking Cradle).

Creative's Zen Vision and Zen Vision:M

Creative's Zen Vision and Zen Vision:M (see Figure 13.13) bring photos and video to the Creative
product line. With a standard 30GB storage capacity, these two devices can store over 120 hours
of video or thousands of songs and photos. They have built-in voice recording and FM receivers,
can output video and stereo audio to television and stereo equipment, and are compatible with
Creative's Digital Video recorders for direct import of video without a computer.

Figure 13.13
The Creative Zen Vision
and Zen Vision:M.

The Vision offers additional connectivity options such as a Compact Flash card slot and a Compact Flash media adapter, wired and wireless remote controls, and high capacity battery options.
The Vision:M is a self-contained unit without the expansion features of the Vision, but offers the
same storage space and media format compatibility as the Vision.

Cables

Creative includes an AV cable with the Vision and Vision:M to let you connect to your television or stereo system. This cable allows the Vision and Vision:M to output composite video and stereo audio.

Compact Flash Card

The Vision includes a Compact Flash card slot that allows you to insert any Compact Flash card for additional storage or media import/export. With Compact Flash cards available in the 8 to 16GB range, you can quickly increase your photo or video capacity. Compact Flash cards are also an excellent way to import media from your digital camera. Shoot the photos on your Compact Flash card and import them directly into the Vision.

Compact Flash Media Adapter

If you camera doesn't support Compact Flash memory, you can use an available Compact Flash media adapter to read five additional memory card formats such as SD and Memory Stick cards.

Remote Control

So, let's say you're watching a movie from your Creative Zen Vision, which is docked and linked to your big screen television. You should pause the movie to answer the phone, but you just sat down and really don't want to drag your tired bones out of the chair again.

Enter the Creative Vision Wireless Infrared Remote Control (see Figure 13.14). This device allows your docked Zen Vision to be controlled remotely just like a DVD player. Start, stop, pause, and rewind your movie. The only thing it can't do is get up and grab the phone for you.

Figure 13.14
Never leave your chair with the Vision IR Remote Control.

265

Mobile Windows Devices

Some Microsoft partners manufacture mobile media players that also double as personal digital assistants. Called Mobile Media Companions by some, they offer a feature set similar to that offered by the Creative Zen Vision series. In addition, they offer the ability to function as remote controls for home entertainment systems, full PDA functionality, and Internet access. Their memory space is more limited than dedicated media devices (rated in megabytes rather than gigabytes), but they are capable of utilizing storage connected to a home network to obtain media from a home computer.

GETTING THE MOST FROM MULTI-FUNCTION DEVICES

What I'm Listening To

Title: Slip Slidin' Away
Artist: Paul Simon
Source: Negotiations and Love Songs (CD)
Player: Windows Media Player 10

The ability to play digital media is being incorporated into a large number of devices. As of the writing of this book I am aware of wristwatches, cameras, telephones, PDAs, even a toilet (see Figure 13.15)! The circuits that play MP3 and other media formats are small, inexpensive, and easy to incorporate into these devices. Manufacturers include them as a quick way to help their product stand out in their marketplace (the toilet also flushes itself and lowers the lid when you leave).

Figure 13.15
The Toto MP3 toilet also flushes itself and closes the lid when you leave the room.

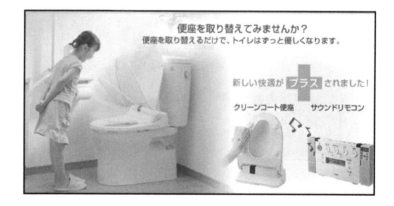

Cameras

A camera that plays MP3s?

Why not?

Creative Labs makes the DiVi CAM 428 (see Figure 13.16), a camera that records digital video, plays MP3s, and shoots 4-megapixel still photos. The DiVi CAM 428—which has very little of its own memory for storage—can link to the Creative Zen Vision to store up to 120 hours of video.

Figure 13.16
The DiVi CAM 428 shoots still photos, video, and plays MP3s.

Telephones

Several telephones are available now that can double as MP3 players. With features such as Internet access, messaging, games, and cameras, the mobile phone is becoming a key device for more than just voice communications.

The iPod ROKR

Motorola and Apple have teamed up to make the ROKR phone, an iTunes compatible telephone with a 100 song capacity. Available only through Cingular in the US, it has met with mixed reviews, especially from Apple fans who expect slick monolithic design. With no Click Wheel and all those buttons (see Figure 13.17), many iPod aficionados find the controls too confusing.

Other Media Phones

Another music phone, "the V" from LG Electronics (see Figure 13.18), plays and records video and MP3 files, and includes the other common mobile phone options such as Internet access, games, musical ring tones, and text messaging. (Oh, and did I mention that you can also make and receive calls with it?)

Figure 13.17
The ROKR iTunes phone
holds about 100 songs.

Figure 13.18
The V phone from LG Elec-
tronics is a digital media
powerhouse.

The Nokia 3250 phone (see Figure 13.19) has storage for up to 750 songs and has the ability to record voice memos, shoot photos, and record video. It also has a full personal information manager with calendar, contacts, and e-mail, a personal networking feature known as Nokia Sensor that is designed to introduce you to other Nokia users in your general vicinity, and built-in blogging software. Wonder where it all fits? Well, Nokia has patented the use of subspace anomalies to store mainframe computers in a small wormhole in the time/space continuum. After that, the rest is simple.

Figure 13.19
The Nokia 3250 phone should have a much bigger case considering all the features they've crammed in.

BE THE LIFE OF THE PARTY!

What I'm Listening To
Title: Invisible Sun Artist: The Police Source: Every Breath You Take CD (WMA   Ripped from CD) Player: Windows Media Player 10

One of the great aspects of digital music is its flexibility. With portable media devices, two or three friends at a party with their iPods or MP3 players probably have a music collection that would shame most professional DJs. All they need is a way to get it into the sound system…

In this section, I'll show you how to use your computer or your portable media player to play the DJ at your next party. I'll also show you how you can let your guests help out at a BYOM (Bring Your Own Music) party.

Windows Media Center as a Jukebox

You've got a few thousand dollars worth of digital music collected over the years. Some ripped from CD, some bought from online music stores. You've got a decent collection of indie music that you've been able to find here and there on artists' websites, and you have just discovered that your significant other has over 3,000 tunes on an iPod. After spending a couple of weekends organizing this mega-collection into a comprehensive music library, you are ready to show off a little.

Playing all this on an iPod connected to a Bose SoundDock would be pretty cool, but you want to really impress all your geek friends. How about letting them call the shots—jukebox-style—with your Windows Media Center computer. You've got the plasma screen hooked up and the computer isn't busy searching for extraterrestrial intelligence this weekend, so why not?

Windows Media Center's Party Mode allows you to designate a playlist for a party and let guests add songs, and create messages to be displayed on the screen. When starting Party Mode, you'll choose an initial playlist and decide whether to let guests access your music library and modify marquee messages (see Figure 13.20).

Figure 13.20
Party Mode setup allows you to choose playlists and security options.

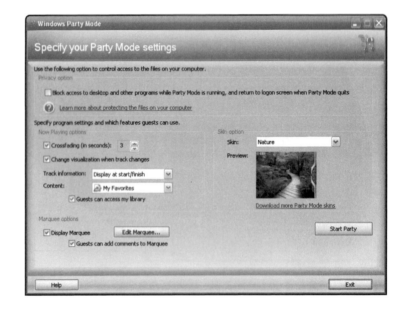

During the party, guests can access your playlist to add songs. Clicking on the plus sign in the playlist display opens the playlist editor (see Figure 13.21). Guests can browse the music library and click on songs to add them to your playlist for the party.

Your Portable Media Device as a Jukebox

Most portable media devices can shuffle a playlist and play it. By connecting your iPod, Zen Micro, iriver PMC, or other portable device to a sound system you can shuffle your playlist and just let the party roll. Not even you will know what the next song will be, so the party will develop its own rhythm. Requests will not be as simple, but the setup is about as simple as it can get. If you do want a little more control, just configure your playlist ahead of time and play it without shuffling.

Figure 13.21
Party Mode playlists can be edited to add requests during the party.

For the Serious Gadget Geek...

If none of the choices I've given you so far are quite geeky enough, just hang on. Here's where we get Cr4z`/ (crazy)!

If you really want to liven up the party, and you have a couple of iPods lying around (who doesn't?), you can try the Numark iDJ (see Figure 13.22). This iPod mixing console mixes the outputs of two iPods and up to two additional players into a single output. Built-in Click Wheels and control buttons let you control the iPods using the built-in menus. Control playback and fade from one device to the other to professionally mix your party. Inputs for microphone and phonograph complete the source options for a totally professional rig.

Professional DJs are excited about this equipment because they can show up at a party with only two iPods, speakers, and this device to replace the cases of CDs and LPs they used to lug around. Hook the iDJ up to a professional sound system and the party will just happen!

> **Tip**
>
> If you don't have two iPods, you can use other portable media players or you can invite a guest DJ to bring another iPod full of tunes.

271

Figure 13.22
The iDJ from Numark is a
DJ's dream.

PUTTING IT ALL TOGETHER (A DAY IN THE LIFE...)

What I'm Listening To

Title: Deora Ar Mo Chroi
Artist: Enya
Source: A Day Without Rain (WMA Ripped
        from CD)
Player: Windows Media Player 10

Okay, you've been with me for 13 long chapters. Let's put this all together and describe a day in the life of a true gadget geek. Obviously, only the most hardcore gadget geek would live like the example I am about to show you. Just consider it a compilation of the ideas presented in this book and not a true account of the actual events in the life of any gadget geek, living or dead.

In other words: Try this only if you are willing to live on Cheetos for the next week as folks pay homage to your divine geekiness.

Dave's Morning Routine

Dave awakened to the light from his Ambient Orb (see Figure 13.23), a device designed to glow different colors based on certain input factors such as outdoor temperature, stock prices, number of unread e-mails, etc. It was a pleasant blue, indicating that Dave's pet chinchilla still had at least an ounce of food left in her dish.

Figure 13.23
The Ambient Orb is a quick visual indicator that can report on a number of trends.

Dave went to the kitchen, his slippers whisper-slapping in the floor as he walked, to see to breakfast. He kicked a few pop cans out of the way and looked into the refrigerator. There were a few slices of cold pizza left from last night's party. That was last night, wasn't it? Oh yeah, the pop cans; it must have been.

Munching on cold pizza and drinking a Red Bull, Dave's eyes started to finally open. The Red Bull kicked in about halfway through the shower and Dave came back out whistling. He ignored the musical greeting from his toilet seat as it opened hopefully at his approach. He would have to calibrate the motion sensor on that. It was opening and playing "Hail to the Chief!" every time he passed and it was beginning to annoy him. He slid into his clothes, cinched up his already knotted tie, grabbed his iPod and his backpack, and headed for the door.

With the iPod docked on his car, Dave listened to weather and news podcasts that had downloaded overnight. When the podcasts finished, he just listened to FM radio for a while.

"How do folks deal with all these commercials?" he mused. He turned the iPod back on and called up his driving playlist. He whistled tunelessly to the music as he finished his commute.

A Day at the Office

Another Ambient Orb on Dave's desk told a very different story about conditions at Nonce Semiconductor, makers of cryptographic equipment. This Orb was glowing an angry red; a sure sign that hackers had been hammering at the firewall again. After a quick check that the intrusion detection system had not sounded an alert, Dave was relieved to see the Orb mellow to a muted orange. There were no stress free days in IT security. Dave couldn't remember the last time this Orb had glowed green.

Meetings and a long lunch took most of the rest of Dave's workday, and he didn't have a chance to catch the streaming broadcast of his favorite security show so he quickly synced the podcast version to his iPod for later. He checked the firewall reports again and, satisfied that the hackers had given up, began applying firewall updates that had recently come back from the quality assurance department.

Dave's cell phone began playing "Kashmir," a signal that Dave's friend Ted, a seventies throwback who still preferred his lava lamp to an Orb, was looking for him. "Yeah?" Dave answered.

"What're you doing tonight?" Ted asked the moment he heard Dave's voice. "I've got *Return of the King* downloaded from CinemaNow. You want to come over?"

"I'd like to, Ted, but Amy's coming over tonight. We're watching one of her DVDs."

"How quaint! A DVD!" Ted groused.

"Shut up! You have at least a dozen of them yourself; more if you count director's commentaries!" (Referring to Ted's *Star Wars* collection.)

"Okay, okay!" Ted grunted a good-bye and hung up.

The Ride Home

Dave left for home, listening to his security podcast in the car. He stopped at the store to pick up a little wine and a couple of TV dinners. He undocked his iPod and wore it into the store so he could keep listening to the show. He wasn't being rude; he just didn't want to miss the show and he knew he wouldn't have a chance to finish it at home.

The TV dinners in the back seat, he headed out onto the freeway. A signboard over the roadway notified drivers that a crash ahead was blocking two lanes of the freeway. Dave exited at the next ramp, calling up a street map on his cell phone. He plotted a route around the crash and began driving across town on surface streets. The security podcast piqued his interest in a particular software application for firewall monitoring so he left himself a quick voice memo in his cell phone to remind himself to check that application out the next day.

An Afternoon Run

He got home only a few minutes later than he normally did. He still had time for a run, so he put the dinners in the freezer, the wine in his electronically controlled wine chiller, and threw on his sweats. He strapped his Creative MuVo NX to his arm (the iPod hard disk just doesn't like the jarring motion of a good run).

He ran his normal route, past Amy's house and down by the river. The barges were plying the river, their push boats churning up the water as they strained to move the massive steel hulks upstream. "Someone's getting an awful lot of coal," he thought to himself. He'd better check out his energy stocks when he got home. His Stock Orb had been green for a while, but you never knew; sometimes it was influenced by blue chips when it should be watching *his* portfolio. He smiled as he wondered if it had a girl Orb on its mind. He recorded a quick memo on his MuVo NX and turned up the hill toward home.

An Evening at Home

The TV dinners were comfortably the same, somewhat bland and tasteless, but better than Dave's own cooking which consisted of various experiments with Cheetos and frozen green beans. Amy said something about him needing someone to take care of him, but Dave completely missed the hopeful innuendo and pointed to his Orbs. "I've got them!" he boasted.

Amy was quiet as they watched the movie together. Wondering what was wrong, but not quite sure how to ask, Dave wished—not for the last time—that he had an Orb for Amy. He decided it had something to do with that "taking care" thing and determined to check out some rings when he had the time. He surreptitiously leaned over and left a quiet memo on his phone, sure Amy couldn't hear his mumbling. He glanced over at her, puzzled by her expression. She seemed to be trying not to laugh or something. Girls! I wish there was an Orb, he thought.

Good Night, Dave!

Dave kissed Amy goodnight, turning back into his apartment. The Chinchilla Orb was turning yellow, so he fed and watered his pet. He stroked her smooth fur thoughtfully, still pondering Amy's sudden mood change. She didn't hear his memo, did she? Does she have a secret mic somewhere in his place?

"Naw! That's silly!" He dropped into bed, soon falling fast asleep.

"Good night, sweet prince." Amy murmured as she drove off.

the gadget geek's guide to

Portable Media Devices

Index

Get Ready to Rock Your Gadget!

Got a new gadget? Use it like a pro right away with these hot new guides!

The Gadget Geek's Guide is a no frills, get-down-to-business series for users who want to make the most of their gadget's features **now**! Each book includes quick tips designed to help you conquer each feature in no time. You'll also get insider advice on techniques used by advanced "techno junkies" who have come up with cool new ways to use the technology. You'll be amazed by what your gadget can do!

The Gadget Geek's Guide to Your BlackBerry and Trēo

ISBN: 1-59863-171-3 ■ $19.99

Also Available